ENVIRONMENTAL HEALTH - PHYSICAL, CHEMICAL AND BIOLOGICAL FACTORS

OIL SPILLS

ENVIRONMENTAL ISSUES, PREVENTION AND ECOLOGICAL IMPACTS

ENVIRONMENTAL HEALTH - PHYSICAL, CHEMICAL AND BIOLOGICAL FACTORS

Additional books in this series can be found on Nova's website under the Series tab.

Additional e-books in this series can be found on Nova's website under the e-book tab.

ENVIRONMENTAL HEALTH - PHYSICAL, CHEMICAL AND BIOLOGICAL FACTORS

OIL SPILLS

ENVIRONMENTAL ISSUES, PREVENTION AND ECOLOGICAL IMPACTS

ADAM CLIFTON
EDITOR

nova publishers
New York

For permission to use material from this book please contact us:
Telephone 631-231-7269; Fax 631-231-8175
Web Site: http://www.novapublishers.com

Library of Congress Cataloging-in-Publication Data

Oil spills (Clifton)
Oil spills : environmental issues, prevention and ecological impacts / [edited by] Adam Clifton.
pages cm. -- (Environmental health, physical, chemical and biological factor)
Includes bibliographical references and index.
ISBN 978-1-63321-515-3 (hardcover)
1. Oil spills--Environmental aspects. 2. Oil spills--Management. I. Clifton, Adam. II. Title.
TD427.P4O38843 2014
628.1'6833--dc23

2014026718

Published by Nova Science Publishers, Inc. † New York

CONTENTS

PREFACE

Petroleum perhaps is one of the most important substances in the modern society due to its wide use in the most diverse productive sectors. Besides serving as raw material to the production of plastics and other compounds, it is the fuel source used in the production of electric energy, in industries and in different transport sectors. However, the ecosystems located in areas of extractions, processing and distribution of petroleum end up being highly susceptible to the direct and indirect impacts of these activities, as well as those located in their transport route. As the large extraction fields are located in coastal areas, these environments are, generally, the most affected, although other ecosystems also suffer influence of petroleum activities. Oil spills and leakages provoke immediate effects into the environment, easily viewed by the generalized mortality of animals and consequent ecosystems losses, but also provoke more persistent effects, such as alteration of the animals behaviour and permanence of compounds derived from petroleum in the food webs. Accidents with petroleum are unpredictable and the resources needed to minimize their impacts, often, are not liable to an immediate application. This book discusses different prevention methods as well as environmental and ecological impacts oil spills have.

Chapter 1 – Immediately after oil spills, assertions often arise that "ecosystems do not recover from spill effects." A survey of the literature concerning the recovery of the biophysical components of terrestrial, freshwater and marine environments assessed the scientific support for the notion of no environmental recovery from oil spills. The survey examined 176 Valued Ecosystem Components (VECs) from 48 oil spills in subarctic and cold temperate regions. The scientific literature on recovery from marine oils spills proved voluminous, whereas that for freshwater and terrestrial spills proved sparser.

The outcomes reported by study authors simply do not support the notion that biophysical environments do not recover. Indeed, the environments and their components quite commonly recover. For the studies reviewed, the VECs had recovered or were recovering by study's end in 83%, 67%, and 72% of the cases for marine, freshwater, and terrestrial biophysical environments, respectively. Despite the lack of clean-up in 10 experimental spills, 62% of the VECs were recovering at study's end. Overall, 79% of the total VECs examined from all environments had recovered or were recovering. Cases of no recovery or no apparent recovery were associated with short study duration, inappropriate spill clean-up techniques, experimental spills, or complex interactions with other natural or human-derived factors.

Recovery times vary with the environment, the VEC, and other factors. The average times to recovered status for the biophysical VECs examined were 1.7 and 4.9 years for freshwater and marine environments, respectively. Recovery times for terrestrial VECs were complicated by the fact that seven of the terrestrial spills were experimental spills in which the oil contamination was not cleaned up. Some terrestrial VECs that achieved recovering status appeared to do so in about 2 years, but others were still recovering or showed no recovery after 20 years, especially among those from experimental spills. Fifteen of the terrestrial VECs were from Arctic and Subarctic Zones where cold temperatures decrease biodegradation rates and short growing seasons decrease annual growth rates.

Several factors can slow recovery. First, if oil persists, recovery times may be lengthened. Clean-up is undertaken to remove and lessen the persistence of hydrocarbons; however, while spill clean-up should strive to remove as much oil as is practically possible, it need not completely eliminate contamination before recovery can proceed. Second, inappropriate clean-up techniques, such as the use of heavy equipment in marshes, can substantially lengthen recovery times. The contingency planning and operations in modern spill response carefully selects the most appropriate treatment for the oil type, level of contamination, and the nature of the receiving environment.

Cases exist where recovery may benefit from active human intervention beyond routine clean-up operations. The most protracted recovery in terrestrial environments appears to be where oil reaches groundwater, where cold temperatures decrease biodegradation rates in soil, or when short thaw seasons limit vegetation growth. The state of the art in bioremediation and phytoremediation continually improves and holds promise for spills where natural environmental processes for oil degradation are inhibited.

Chapter 2 – Much attention has been drawn to oil-degrading microorganisms in the Gulf of Mexico region in order to bioremediate the oil spills after the Deepwater Horizon accident in 2010, while little has been investigated in the impact of spilled oils on microbial pathogens in the Gulf region. This chapter summarizes the observations and analyses of microbial pathogens impacted by the Deepwater Horizon oil spill in the Gulf of Mexico region. In the observation section, microbial pathogens and their waterborne disease outbreaks in the United States are briefly reviewed. Especially in this section, two types of water are focused; tap water treatment procedure in relation to waterborne pathogens and recreational water as a route of disease transmission. Then, waterborne disease outbreaks after natural or anthropogenic disasters and abiotic factors affecting the fate of waterborne pathogens are discussed. At the end of this section, recent publications on the microbial pathogens in the Gulf region after the Deepwater Horizon oil spill are introduced, which are mainly focused on pathogenic marine *Vibrio* species that cause seafood poisonings and skin infections. In the analyses section, various molecular techniques are introduced to detect and monitor marine microbial pathogens including bacteria, protists, and viruses. Lastly, a brief chronicles of the U.S. government regulations to control waterborne disease outbreaks in the Gulf region are introduced and several aspects to control future pathogen outbreaks in the Gulf region are discussed.

Chapter 3 – In this chapter, the authors present a state-of-the-art system developed to manage oil spills crisis at the coastal areas of the Balearic Islands Archipelago. Specifically, they introduce a first version of an Integrated Decision Support System (IDSS) to response against oil spill emergencies which combines a numerical tool for tracking the spill (or objects) at the sea surface, i.e., the physical-geomorphological component, with a GIS tool containing the environmental and human variables. This system provides in real time the areas of probability of a spill the potential impacts at the shore, the uses of the affected coasts and the resources available in the surrounding areas.

Chapter 4 – Petroleum perhaps is one of the most important substances to the modern society, due to its wide use in the most diverse productive sectors. Besides serving as raw material to the production of plastics and other compounds, it is the fuel source used in the production of electric energy, in industries and in the different transport sectors. However, the ecosystems located in areas of extractions, processing and distribution of petroleum end up being highly susceptible to the direct and indirect impacts of these activities, as well as those located in their transport route. As the large extraction fields are located in coastal areas, these environments are, generally, the most affected, although other ecosystems also suffer influence of the petroleum activities. Oils spills and leakages provoke immediate effects into the environment, easily viewed by the generalized mortality of animals and consequent ecosystems losses, but also provoke more persistent effects, such as alteration of the animals behaviour and permanence of compounds derived from petroleum in the food webs. Accidents with petroleum are unpredictable and the resources needed to minimize their impacts, often, are not liable to an immediate application. Among the strategies developed to recover the impacted environments, there are the mechanical cleaning methods, use of chemical dispersant products and use of bioremediation treatments. Cleaning methods of areas contaminated by petroleum are only considered effective when they prevent the migration of this pollutant to more sensitive ecosystems. Now, the bioremediation has been considered a "green" alternative for treatment of dangerous contaminants since it uses organisms naturally able to reduce, remove and transform the organic and inorganic pollutants. The success of this process depends on it natural biodegradability, bioavailability and optimization of the biological activities related to the pollutant. Among the strategies available to perform the natural bioremediation there are the natural attenuation, bioventilation, biomagnification, biostimlation, landfarming, composting and phytoremediation. In this chapter it will be discussed the concepts related to the impacts caused by petroleum and its derivatives, as well as the ecological consequences of some petroleum accidents and the efficiency of the application of the environmental decontamination techniques, such as bioremediation, since this is an effective strategy, economically viable and that provides a lower environmental impact.

Chapter 5 – During oil spill episodes, the oil may reach the coastal environments, despite efforts to retain and remove the oil from the water. Thus, response actions must be properly planned in order to minimize the environmental impacts caused by the oil. Contingency plans for oil spill accidents were developed in the 1970s and involve a set of steps that include environmental sensitivity index maps (ESI maps). At that time, this valuable tool represented an advancement in action planning. Since then, however, contingency plans have used the ESI maps as a definitive planning solution, without combining them with other tools such as numerical modeling. The technological advances involved in numerical modeling, computational and geo-processing techniques, and access to a wide database through the internet, all allow for the introduction of advancements in oil spill response action plans. This chapter aims to present some of these new techniques in order to improve the classic approaches and provide more efficiency for emergency action plans. One of these techniques is the combination of ESI maps with information from numerical modeling of oil slicks and the probability that a given environment will be reached by the oil. This combination helps to generate maps of the indices of environmental vulnerability to oil (IEVO maps), which quickly and easily show areas that should receive priority protection after an oil spill. Another planning tool that may be used with the ESI and IEVO maps is the map of cleaning procedures and response actions. This map relies on the use of pictographic icons that represent the cleaning procedures recommended for each coastal environment. A careful definition of the recommended methods may keep the response actions from amplifying the damage caused by the oil, and their representation as icons on a map also allows for rapid visualization, and thus rapid decision making by officials responsible for emergency actions. These maps must also rely on solid information; thus, the assessment of biological resources of each coastal environment needs to be complete and deep. The use of GIS techniques to spatially display the biological information may help to identify more sensitive areas within a region that has already been classified using a given environmental sensitivity index. ESI maps that prioritize the representation of biological resources along the potentially affected areas provide advancement in terms of the protection and mitigation of impacts on coastal ecosystems. These new approaches allow for better emergency planning move the classic approach forward, thus contributing to the mitigation of potential environmental impacts caused by oil spills.

Chapter 6 – Environmental tracers, as isotopes, have been used to reveal origin, movement and dispersion of compounds because they behave the same as environmental processes. These isotopes are useful tools for understanding dynamics of natural systems, providing a unique tool to trace the biogeochemical cycles within ecosystems. The hydrological complex of Babitonga Bay (Brazil) forms a vast environmental system; hosting the last great expanse of mangrove forests in the southern hemisphere. Mangroves are among the most productive ecosystems on earth. Effects of an oil spill on mangrove ecosystems were studied in Babitonga Bay using lead and carbon isotopes. Samples of the spilled oil, as well as sediment and water samples, were obtained nine months after the accident, at the time of the salvage operation. Isotopic composition of the oil was utilized to trace the extent of the environmental pollution. A In addition, isotopes of lead and carbon allowed for the identification of areas where oil was present. Contaminated sediments exhibited an isotopic composition (^{206}Pb/^{207}Pb and ^{208}Pb/^{206}Pb) close to that of the oil spilled. δ^{13}C data confirmed these results. The results of this investigation suggest that lead isotope ratios can be very useful in the field of environmental forensics.

Chapter 7 – Crude oil is a mixture of many toxic compounds, including some known mutagens and metals such as polycyclic aromatic hydrocarbons and lead, respectively. The steps of exploration, onshore drilling, transportation and refining involve risk of accidental spills and may compromise environmental health and wildlife inhabiting the area. Among the animal species, free-living rodents play an important role either as seed dispersers or prey to many predators, since they represent intermediate links in the food web. These animals have been commonly used as indicators of disorders caused by environmental contaminants, especially metallic elements, and more recently organic contaminants, such as crude oil. The authors discuss the use of biomarkers in two species of free-living rodents exposed to simulated onshore oil spills, the Algerian mouse *Mus spretus* and the small Vesper mouse *Calomys laucha*. The studies indicate that genetic, hematological, physiological and behavioral biomarkers may be useful tools to predict stress situations facing a condition of an oil spill and point out those wild rodents could be used as risk indicators in natural populations exposed to this environmental contaminant.

Chapter 8 – Oil spills occur either in the ocean and coastal waters or on land due to the release of crude oil from tankers, offshore platforms, drilling rigs and wells. It is becoming a major widespread environmental problem, especially in coastal areas subjected to high anthropogenic inputs. The high crude oil concentrations accumulated in surface and subsurface of ocean and soil pose serious concerns for the environmental and human health. Bioremediation, mainly by indigenous bacteria, has been regarded as an effective way to clean up oil pollution after oil spills. In order to obtain a systematic understanding of crude oil bioremediation, two biodegradation approaches, bio-stimulation and bio-augmentation will be summarized. Moreover, the detailed studies of the Deepwater Horizon oil spill in the Gulf of Mexico and the Penglai 19-3 oil spill in the Bohai Bay, China will be compared. The immobilization techniques used in the bioremediation process will also be reviewed. The insights and theories of these studies will improve our understanding of the microbial degradation process and mechanisms of indigenous microorganisms and provide fundamental evidences and strategies for *in situ* bioremediation of crude oil spills in the marine ecosystem.

Chapter 9 – Oil spill often leads to significant, long-term impacts on the environment. This requires a best decision to be made at the right moment. Hence, evaluating combat options in a sufficient way is still an ongoing concern challenging decision makers. A set of analysis tools including multi-criteria analysis (MCA), fuzzy comprehensive evaluation (FCE) and monetary evaluation model (MEM) have been developed and applied recently in the management for oil spill contingency. Following a close comparison of the methods, selections of evaluation methods are finally summarized for development of good practice in the context of integrated coastal zone management (ICZM).

In: Oil Spills
Editor: Adam Clifton

ISBN: 978-1-63321-515-3

Chapter 1

BIOPHYSICAL ENVIRONMENTS DO RECOVER FROM OIL SPILL EFFECTS

Walter H. Pearson[1,*] and W. Cody Fleece[2,†]
[1]Stantec Consulting Services, Suite E, Okemos, MI, US
[2]Stantec Consulting Services, Cincinnati, OH, US

ABSTRACT

Immediately after oil spills, assertions often arise that "ecosystems do not recover from spill effects." A survey of the literature concerning the recovery of the biophysical components of terrestrial, freshwater and marine environments assessed the scientific support for the notion of no environmental recovery from oil spills. The survey examined 176 Valued Ecosystem Components (VECs) from 48 oil spills in subarctic and cold temperate regions. The scientific literature on recovery from marine oils spills proved voluminous, whereas that for freshwater and terrestrial spills proved sparser.

The outcomes reported by study authors simply do not support the notion that biophysical environments do not recover. Indeed, the environments and their components quite commonly recover. For the studies reviewed, the VECs had recovered or were recovering by study's end in 83%, 67%, and 72% of the cases for marine, freshwater, and terrestrial biophysical environments, respectively. Despite the lack of clean-up in 10 experimental spills, 62% of the VECs were recovering at study's end. Overall, 79% of the total VECs examined from all environments had recovered or were recovering. Cases of no recovery or no apparent recovery were associated with short study duration, inappropriate spill clean-up techniques, experimental spills, or complex interactions with other natural or human-derived factors.

Recovery times vary with the environment, the VEC, and other factors. The average times to recovered status for the biophysical VECs examined were 1.7 and 4.9 years for freshwater and marine environments, respectively. Recovery times for terrestrial VECs were complicated by the fact that seven of the terrestrial spills were experimental spills in which the oil contamination was not cleaned up. Some terrestrial VECs that achieved recovering status appeared to do so in about 2 years, but others were still recovering or

* Phone: 616 -610-4614. E-mail: Walter.Pearson@stantec.com.
† E-mail: Cody.Fleece@stantec.com.

showed no recovery after 20 years, especially among those from experimental spills. Fifteen of the terrestrial VECs were from Arctic and Subarctic Zones where cold temperatures decrease biodegradation rates and short growing seasons decrease annual growth rates.

Several factors can slow recovery. First, if oil persists, recovery times may be lengthened. Clean-up is undertaken to remove and lessen the persistence of hydrocarbons; however, while spill clean-up should strive to remove as much oil as is practically possible, it need not completely eliminate contamination before recovery can proceed. Second, inappropriate clean-up techniques, such as the use of heavy equipment in marshes, can substantially lengthen recovery times. The contingency planning and operations in modern spill response carefully selects the most appropriate treatment for the oil type, level of contamination, and the nature of the receiving environment.

Cases exist where recovery may benefit from active human intervention beyond routine clean-up operations. The most protracted recovery in terrestrial environments appears to be where oil reaches groundwater, where cold temperatures decrease biodegradation rates in soil, or when short thaw seasons limit vegetation growth. The state of the art in bioremediation and phytoremediation continually improves and holds promise for spills where natural environmental processes for oil degradation are inhibited.

INTRODUCTION

Immediately after oil spills, assertions often arise that "ecosystems just do not recover from spill effects." This notion needs to be confronted with the results of scientific studies of recovery. The purpose of this chapter is to examine what the scientific literature has to say concerning the recovery of the biophysical components of terrestrial, freshwater and marine environments from oil spills. This chapter reports the results of our most recent examination of the scientific literature regarding recovery of ecosystems and their Valued Ecosystem Components (VECs) from spills that have occurred primarily in Cold Temperate and Subarctic Zones. The chapter addresses two sets of questions.

The first set of questions includes:

- How do oil spill characteristics differ among recovery studies for the terrestrial, freshwater, and marine environments?
- How common is recovery?
- How much time does recovery take?
- What factors influence recovery?

A semi-quantitative analysis of our results from the literature survey addresses the above questions.

The second set of questions includes:

- How does recovery from oil spills compare to recovery from other disturbances?
- What factors impede recovery?
- What can be done when recovery lags?
- What implications do the survey results have for oil spill preparedness, prevention and response?

- What implications do the survey results have for defining and measuring recovery?

A literature review approach addresses this second set of questions and provides additional information concerning questions from the first set. Although clean-up is usually the first step to environmental recovery, this chapter examines the course of events after the clean-up phase has ended to assess recovery processes and outcomes for the biophysical environment.

Many scientific papers and books have reviewed oil spills (Clark 1982; Teal and Howarth 1984; Wells et al. 1995; Rice 1996; API 1999; National Research Council 1975, 1985, 2003; Kingston 2002; Wiens 2013), but here we focus on those studies that have followed recovery. Recovery of human environments is not treated here but one example of a study documenting recovery of the human environment from an oil spill can be found in Fall et al. (2006).

This chapter represents the third iteration in our ongoing study of recovery. The first and second iterations were prompted by hearings conducted as part of a review for the proposed Northern Gateway Pipeline between Alberta and British Columbia, Canada. The Canadian National Energy Board and the federal Minister of the Environment established a Joint Review Panel to conduct a review of the project. The specific objective of the initial review was to assess the scientific evidence regarding ecological recovery from oil spills. Initially, about 114 papers were reviewed for the first iteration (Stantec 2012) and assessed for relevance to recovery in support of the hearing process. This chapter is now based on reviews of over 140 publications and builds upon the original report (Stantec 2012) by incorporating information and data developed during the hearings and from subsequent review of the literature following the close of the regulatory hearings. The additional literature corroborates the results of the initial review and lends stronger support to the major conclusions of our initial review.

This introduction includes two short overviews drawn from the literature. The first addresses the behavior of spilled oil, and the second, the various processes involved in recovery.

Behavior of Spilled Oil in the Environment

The behavior of spilled oil is governed by the characteristics of the environment at the time and place of the spill and by the characteristics of the spilled oil, including type and the rate and duration of spillage. Substantially greater detail appears in reviews by the National Research Council and others (NRC 1975, 1985, 1999, 2003, Whittle et al. 1982, Wiens 2013). The general framework for understanding oil behavior derives primarily from marine spills of crude oil and some refined products, which are the most studied overall. Some observations concerning spilled oil behavior in terrestrial and freshwater environments appear at this section's end.

Crude oils and refined products such as fuel oils are mixtures of natural chemicals. The modifications to crude oil in refining can significantly alter the products' behavior and fate in the environment, but the petroleum-based aromatic and aliphatic compounds in the refined products undergo the same degradation processes as the original parent compounds.

Immediately upon being spilled into terrestrial, freshwater, and marine environments, the spilled oil becomes subject to numerous chemical, physical and biological processes that begin to break down, biodegrade and otherwise assimilate the spilled oil (NRC 1975, 1985, 1999, 2003, Whittle et al. 1982). This natural degradation of oil sets the conditions under which the recovery of the biophysical and human environments from oil spills occurs. Ultimately, spilled oil is broken down into carbon dioxide and water by sunlight (photolysis) and microbes (biodegradation). However, the time required depends on the nature and amount of the petroleum product spilled and the characteristics of the receiving environment.

Oil spilled onto water surfaces under calm conditions tends to spread by gravity into a thin layer (about 0.1 mm thick) to cover a relatively large area. However, the actions of wind, tides, and currents often lead to an uneven distribution of oil slicks of various thicknesses. Wind and current can transport the slicks, which then can strand on shorelines. Oil viscosity increases as the slicks weather, and spreading decreases. Under windy conditions, many oils form stable water-in-oil emulsions called mousse - the sticky, dark coagulum that can accumulate along shorelines. Because water becomes incorporated into the mousse, the amount of contaminated material that must be addressed in the spill response also increases. Strong winds increase the rate of evaporation and, along with wave action, can increase the rate of dissolution. Both weathering processes remove the more volatile and soluble hydrocarbons from the oil.

Different weathering processes dominate at different times after spillage. Within hours evaporation transports most of the volatile and toxic oil components into the atmosphere, where sunlight oxidizes the compounds. Evaporation can remove up to 50% of the oil mass from the water surface, depending on oil type and its content of volatile compounds. Dispersion occurs when wind waves or other turbulence drives oil into the water column where it is present as droplets or dissolved hydrocarbon for the more soluble fractions. Oil droplets can coalesce into larger droplets that float back to the water surface when the waves or turbulence abates.

The terms "dispersed," "sunken," and "submerged" oil are frequently used interchangeably but are not precisely the same because different processes are involved (NRC 1999). Naturally dispersed oil occurs when turbulence drives oil into the water column. Such conditions as moderate waves in marine and lake environments or high, turbulent flows in rivers can produce this turbulence. Dispersed oil is present as droplets that can coalesce and resurface when the turbulence decreases. Submerged oil can occur when the oil density is greater than that of the receiving water and the oil sinks by gravity. Increased oil density can occur when weathering removes the lighter volatile compounds or, more commonly, when oil comes into contact with sand and incorporates heavy sand particles. Distinguishing the mechanisms that have operated in a given spill can be difficult so that NRC (1999) recommends using the term "non-floating oil."

Oil spilled into the terrestrial environment spreads by gravity but the spreading rate and pathway is heavily influenced by the terrain, soil types, and vegetation. The tendency of oil to seep into soil and migrate into groundwater is affected by soil characteristics such as grain size and by oil characteristics such as viscosity and weathering state. Oil spreading over land tends to pool in low areas in the terrain. Oil spilled on land can cross the terrain to reach freshwater. In crossing the terrain, spilled oil can incorporate heavy particles.

Oil spilled into freshwater systems, such as streams, rivers, ponds, and lakes, is subject to the same processes of evaporation, dispersion, dilution, dissolution, and transport as oil in marine systems (API 1999, NRC 1999). As noted earlier, the turbulence associated with swift currents in streams and rivers can physically disperse oil into the water column. As in marine systems the dispersed oil droplets can coalesce when the turbulence decreases and the larger droplets float to water surface. Because of the shallow depths of rivers compared to marine waters, there is greater opportunity for the dispersed oil to reach bottom substrates and sorb particulate matter. Freshwater is less dense than marine waters and the sorption of particulates can increase the density of heavy oils to the point that it sinks by gravity.

Ultimately, most of the oil spilled into terrestrial, freshwater and marine environments is broken down by natural agents. With no clean-up or other treatment, about 75% of crude oil spilled into the marine environment will be broken into carbon dioxide and water (Whittle et al. 1982). Biodegradation rates depend on the oil type and characteristics of the receiving environment, such as temperature and prevailing microbial populations (NRC 2003).

Processes in Recovery

First, some distinctions concerning terminology are needed. The term "oil spill recovery" can be ambiguous. Here, the term "recovery" refers to the recovery of the biophysical environment from an oil spill, not the recovery or collection of oil during oil spill response and clean-up. The physical collection and removal of oil from the environment is an important first step in enabling recovery and decreasing time to recovery but is not recovery in the sense used here. Also, remediation and restoration are active human interventions that enable or accelerate recovery, but are not recovery itself. Ecological recovery, then, includes the processes and outcomes that return the ecosystem or VEC to a desired functioning state.

Some natural processes that advance ecological recovery include:

- Natural transport – wave action, currents, and other processes physically move the oil from one environmental compartment to another;
- Natural transformation – photolysis and biodegradation break down the oil;
- Natural population growth – the population rebuilds through natural recruitment and growth;
- Immigration, colonization, and recolonization – organisms move into the affected area and re-establish their populations.

This chapter considers the factors in recovery in more detail later.

Approach

The study objective was to examine published studies to discern the broad patterns in ecological recovery following oil spills. Accomplishing the objective required three steps: 1) identification of spills where ecological recovery was studied 2) construction of a database of attributes described in those studies (Appendix), and 3) analysis of the data for patterns regarding ecological recovery.

Spill Selection

Candidate spills were identified for inclusion in the study using widely available databases including EBSCOhost, ScienceDirect, and Google Scholar. Articles were also identified from the bibliographies of relevant articles. The selection process for literature was not an attempt to randomly sample all oil spills but rather to obtain a list of spills most relevant to recovery of the biophysical environment in Cold Temperate and Subarctic Zones. General criteria for inclusion of literature for a spill were the following:

- Only studies that had some explicit mention of recovery were examined in detail and included in the database for assessment.
- Studies examined were those that documented changes in conditions of a particular VEC with time after an oil spill.
- Studies were from cold temperate, subarctic, and temperate ecoregions as a means to narrow the assessment to circumstances most relevant to the Northern Gateway project.
- In a small number of cases, studies were included in the assessment despite having failed to satisfy these criteria if the spill represented the only source of information for a particular topic area.

The largest and most intensively studied oil spills often focused on more than one Valued Ecosystem Component (VEC) while smaller or less intensively studied spills might report on only one VEC. Nine of the 48 spills in the database accounted for 116 VECs while the remaining 39 spills had three or fewer VECs. The majority of the spills examined were from the Northern Hemisphere (Figure 1).

Information Sought

The following variables were initially identified for inclusion in the database.

- Oil spill name
- Location
- Year of spill
- Oil type
- Volume spilled
- Volume recovered
- Platform (source) for spill
- Environment affected
- Valued Ecosystem Component
- Observed mortality
- Recovery status
- Processes reported
- Measures to accelerate recovery
- Years to recovery
- Study duration

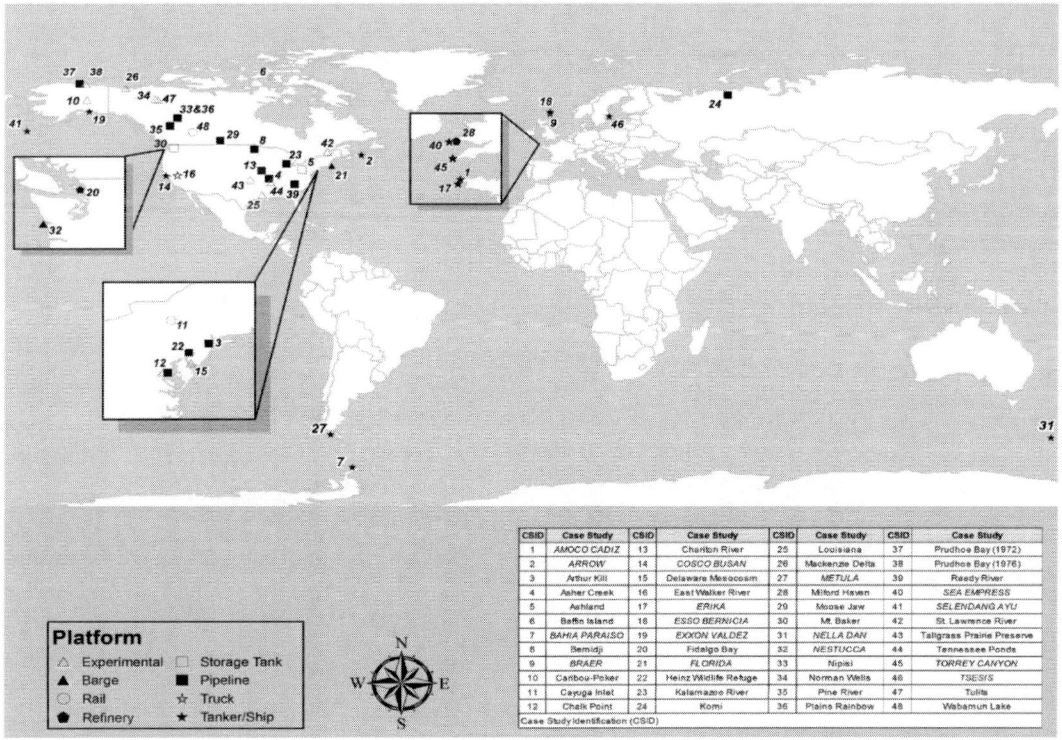

CSID	Case Study	CSID	Case Study	CSID	Case Study	CSID	Case Study
1	*AMOCO CADIZ*	13	Chariton River	25	Louisiana	37	Prudhoe Bay (1972)
2	*ARROW*	14	*COSCO BUSAN*	26	Mackenzie Delta	38	Prudhoe Bay (1976)
3	Arthur Kill	15	Delaware Mesocosm	27	*METULA*	39	Reedy River
4	Asher Creek	16	East Walker River	28	Milford Haven	40	*SEA EMPRESS*
5	Ashland	17	*ERIKA*	29	Moose Jaw	41	*SELENDANG AYU*
6	Baffin Island	18	*ESSO BERNICIA*	30	Mt. Baker	42	St. Lawrence River
7	*BAHIA PARAISO*	19	*EXXON VALDEZ*	31	*NELLA DAN*	43	Tallgrass Prairie Preserve
8	Bemidji	20	Fidalgo Bay	32	*NESTUCCA*	44	Tennessee Ponds
9	*BRAER*	21	*FLORIDA*	33	Nipisi	45	*TORREY CANYON*
10	Caribou-Poker	22	Heinz Wildlife Refuge	34	Norman Wells	46	*TSESIS*
11	Cayuga Inlet	23	Kalamazoo River	35	Pine River	47	Tulita
12	Chalk Point	24	Komi	36	Plains Rainbow	48	Wabamun Lake

Case Study Identification (CSID)

Figure 1. Geographic distribution and platform for spills examined in this study.

Very few studies contained comprehensive or complete information for all of the variables of interest. Consequently, sample size varies depending on the completeness of the variables under consideration. Several variables were only sporadically reported in published studies and sample sizes were insufficient for analysis. Variables with insufficient information included volume recovered, observed mortality, processes reported, and measures to accelerate recovery.

Spill Characteristics

This chapter addresses information from a total of 48 spills. Reviewers recorded information on the spill name, location, the year of spill, volume, and type of product spilled.

Spilled products were described in the literature as crude oil, heavy crude, light crude, Number 2 Fuel Oil, Number 5 Fuel Oil, Number 6 Fuel Oil, Diesel, gasoline, and various mixtures of the aforementioned types (i.e., multiple products were spilled). The product types were consolidated according to shared properties into the following categories: crude oil, fuel oil, or crude and fuel oil. Only one gasoline spill, which accounted for only one VEC, was included in the initial database. This record was omitted from consideration here (Appendix) because of the small sample size and the different properties of gasoline compared to crude and fuel oils. Recovery was studied in accidental spills originating from several platforms (e.g., tankers, pipelines, trains), as well as in experimental studies in which oil was intentionally released or left untreated.

Environmental Characteristics

Spills were classified as Marine, Terrestrial, or Freshwater according to the environment where the spill occurred. Within these groups, spills were further classified according to the Valued Ecosystem Component affected and for which recovery was studied. The Marine category included estuaries. Information was assembled for 176 VECs grouped into the following categories:

- Algae
- Birds
- Fish
- Macro-invertebrates
- Mammals
- Microbes
- Reptiles
- Vegetation
- Sediment
- Shoreline
- Soil
- Water Quality

VECs in the transition zone between aquatic and terrestrial transition environments (e.g., shorelines and shoreline vegetation) were included in Marine and Freshwater categories. Upland vegetation, soil and groundwater VECs were assigned to Terrestrial ecosystems. VECs most often focused on single species but studies also focused on community or assemblage metrics (e.g., diversity, richness, Index of Biotic Integrity).

In some instances, the same VECs were studied at multiple points in time. For example, Collins et al. (1994) studied product constituents in soils 15 years after an experimental spill and Prince et al. (2003) analyzed product constituents in soils at the same location 10 years later (25 years post-spill). Therefore, although some VECs may appear to be double counted, multiple studies on the same VEC for the same spill provide information on recovery over time.

Recovery Status

A prerequisite of the recovery evaluation was that sufficient information for each VEC existed that would allow for the detection of injury and for tracking progress toward recovery. Each paper was evaluated for recovery in the form of quantitative data showing temporal change or simply using general keywords such as "recover (or any part of the word), return to natural equilibrium, return to previous historical conditions, return to pristine conditions, and return to baseline." If substantive data were presented characterizing the status of a VEC, reviewers accepted the author's conclusions regarding recovery as presented in the paper without assessment of the validity or accuracy of the conclusions.

The Categories of recovery status and criteria used to guide decisions regarding recovery status were the following:

- Insufficient Data: the author stated that insufficient information was found to judge recovery; the author mentioned recovery but did not present sufficient information or data for the reader to discern recovery status; the author presented an assertion concerning recovery or the absence of recovery but offered no data or chain of reasoning to support the conclusion;
- None or none apparent: not showing recovery, long term effects have set in and conditions are not improving or improve very modestly;
- Recovered: full recovery or a return to pre-spill conditions that were equal to non-oil spill base data or baseline conditions;
- Recovering: partial recovery, moving closer to conditions before oil spill, but not yet exhibiting full recovery;
- Uncertain: this category refers to circumstances where some metrics, sample locations, or time periods suggest that some recovery has occurred, while other indicators suggest that recovery has not occurred.

Some spills were studied at multiple points in time. Reviewers tracked each spill through to the most recent, readily available publication. However, it was often observed that when resources were categorized as "recovering," no additional studies could be found verifying "recovered."

Assessment and Analysis

Simple descriptive statistics were generated from the assembled database (Appendix) characterizing the type and number of spills encountered in the literature, the duration of study by ecosystem, the number and volume of spills by environment and other appropriate indicators. A total of 176 VECs from 48 spills formed the base of this analysis. In a few cases, correlation or association tests were used to examine relationships between variables.

RESULTS

How Do Oil Spill and Other Characteristics Differ Among Recovery Studies?

Study of recovery from spills was quite uneven among environments (Figure 2a) and VECs (Figure 2b). The spill characteristics in this chapter do not represent those for all spills, but for those for which some study of recovery was found.

Marine spills dominate with the most VECs studied and the most spills with study durations greater than 5 years (Table 1 and Figure 3). Less attention has been given to recovery in freshwater and terrestrial environments. Similarly, macroinvertebrates, shorelines, birds, sediment, and fish have received the most study of recovery, while the rest substantially less (Figure 4).

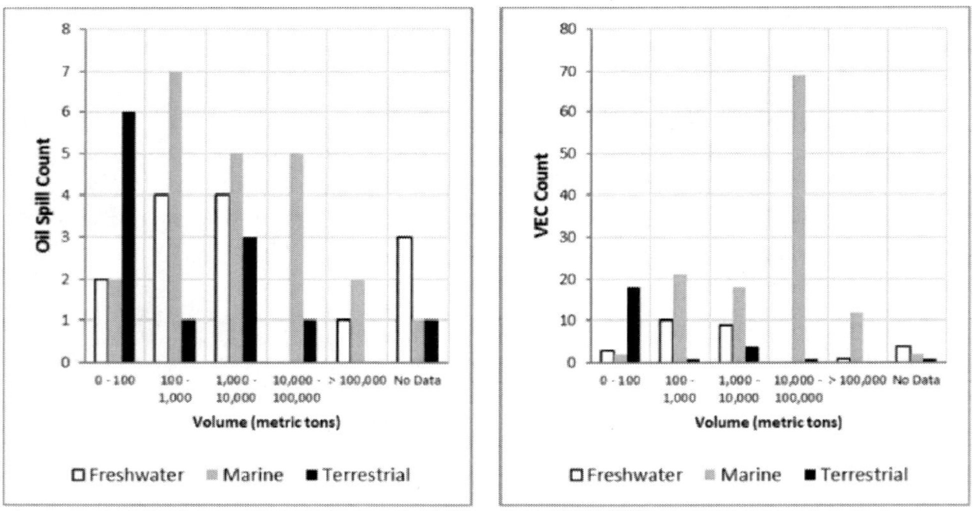

a b

Figure 2. (a) Spill frequency by environment and volume and (b) VEC frequency by environment and spill volume.

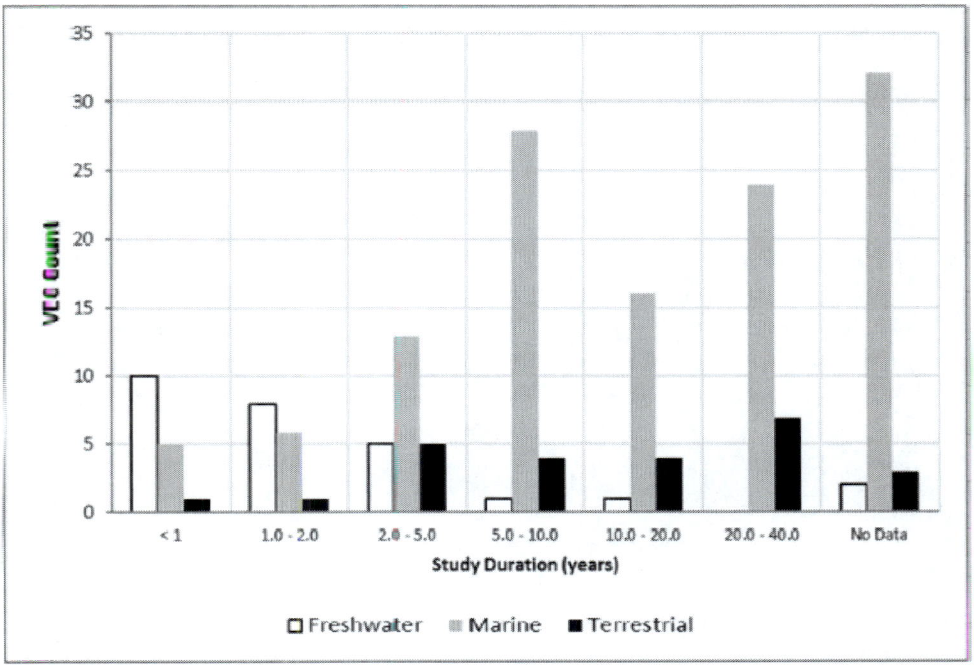

Figure 3. VEC count by environment and study duration.

Crude oils were by far the most common oil type for which recovery was studied (Figure 5). Fuel oils of various types were the second most common oil types studied. Some spills involved release of both crude oil from the cargo and a fuel oil from the fuel tanks.

Crude oils have been studied in all three environments and crude oil spills from tanker ships have received the most attention (Figures 5 and 6 and Table 2). The largest spill volumes were also associated with tankers and ships (Table 2).

Table 1. Spill volume and study duration by environment

	Volume (metric tons)			Study Duration (years)		
Environment	Spill Count	Average	Standard Deviation	Count	Average	Standard Deviation
Freshwater	11	13434	40790	25	1.9	2.8
Marine	21	30839	58641	92	12.1	9.9
Terrestrial	11	8785	25245	22	12.9	10.6
Grand Total	43	20745	47831	139	10.4	10.0

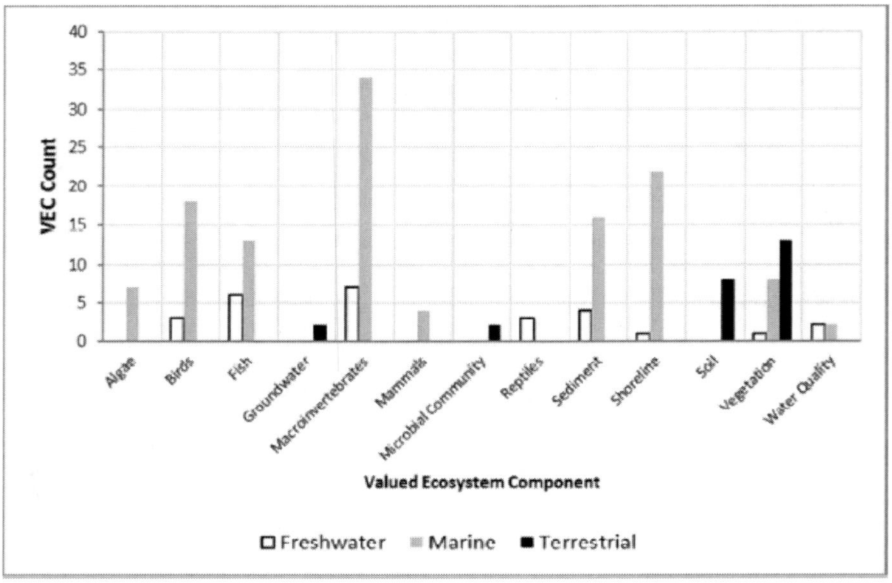

Figure 4. VEC count by VEC group and environment.

Terrestrial spills reported in the recovery literature tend to be from pipelines, refinery storage tanks, and railway accidents. Freshwater spills are primarily from barges and pipelines. Most VECs came from large marine spills (Figure 6) and the EXXON VALDEZ oil spill (EVOS) had the largest VEC count (n=54) of any spill.

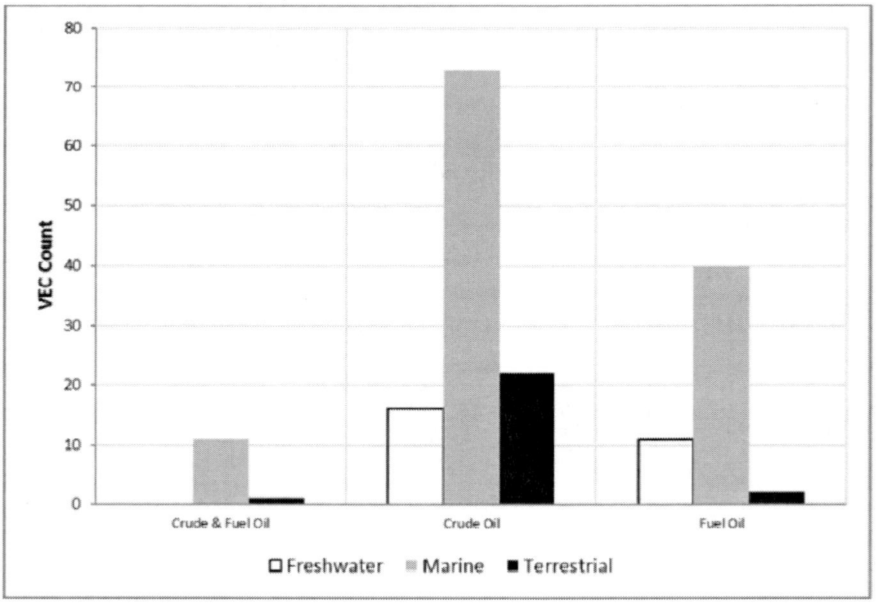

Figure 5. VEC count by product type and environment.

How Common Is Recovery?

The sum of the study outcomes as reported by study authors simply does not support the notion that ecosystems do not recover from oil spills (Table 3, Figure 7).

Indeed, not only do ecosystem components recover, and they do so quite commonly. For the 48 oil spills reviewed, the VECs had recovered or were recovering by study's end in 83%, 67%, and 72% of the cases for marine, freshwater, and terrestrial biophysical environments (n=124, 27, and 25), respectively. The ten experimental spills had 21 VECs subject to either intentional release or intentional lack of clean-up. Despite this lack of clean-up, 62% were recovering at study's end (Appendix). Overall, 79% of the total VECs examined from all environments and platforms (n=176) had recovered or were recovering.

Table 2. Spill volume and study duration by spill source

	Volume (metric tons)			Study Duration (years)		
Platform	Spill Count	Average	Standard Deviation	Count	Average	Standard Deviation
Barge	2	646	125	10	19.9	14.9
Experimental	7	12121	31938	19	8.9	9.2
Pipeline	14	11560	35991	24	8.2	9.8
Rail	2	75	73	6	0.9	0.7
Refinery	1	670	NA	4	11.7	12.0
Storage Tank	2	1128	1564	3	3.7	2.9
Tanker/Ship	14	45785	67585	71	11.4	8.8
Truck	1	11	NA	2	0.4	0.0
Grand Total	43	20745	47831	139	10.4	10.0

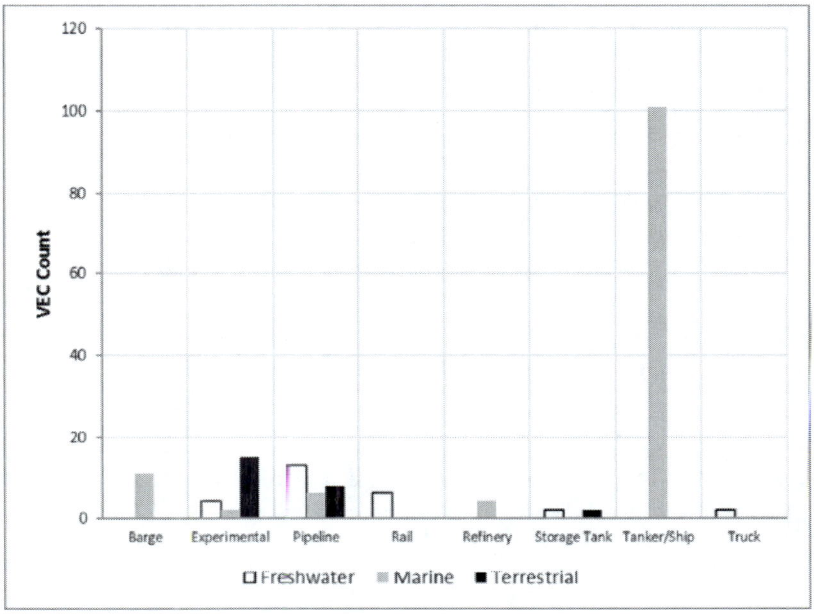

Figure 6. VEC count by spill source and environment.

The frequency of recovering VECs was higher in fuel oil spills than in crude oil spills relative to other recovery states (Figure 8a). However, these differences were not significant at the 0.05 significance level (χ^2 = 4.969, df = 2, p = 0.083). Observed frequencies for recovery states in freshwater and marine environments (Figure 8a) did not differ significantly (χ^2 = 3.915, df = 2, p = 0.141).

Spills with insufficient information were primarily those with study durations less than 2 years (Table 3). VECs showed no recovery or no apparent recovery in 6%, 11%, and 28% of the cases for marine, freshwater, and terrestrial spills, respectively.

Table 3. Recovery status by environment and study duration

	Proportion			Study Duration			Years to Recover		
Recovery State	VEC Count	Proportion	95% Confidence Interval	Count	Mean	95% Confidence Interval	Count	Mean	95% Confidence Interval
Insufficient Data	4	0.15	0.13	3	1.1	1.0			
None or none apparent	3	0.11	0.12	3	1.3	1.7			
Recovered	5	0.19	0.15	5	2.9	3.2	5	1.7	1.4
Recovering	13	0.48	0.19	12	2.0	1.8			
Uncertain	2	0.07	0.10	2	0.3	0.4			
Freshwater Total	27	0.15		25	1.9				
Insufficient Data	4	0.03	0.03	4	1.3	1.2			
None or none apparent	8	0.06	0.04	7	14.6	6.8			
Recovered	51	0.41	0.09	28	12.3	2.7	48	4.9	1.4
Recovering	52	0.42	0.09	46	13.3	3.3			
Uncertain	9	0.07	0.05	7	6.9	5.1			
Marine Total	124	0.70		92	12.1				
Insufficient Data	0			0					
None or none apparent	7	0.28	0.18	6	9.6	6.2			
Recovering	18	0.72	0.18	16	14.1	5.6			
Recovered	0			0					
Uncertain	0			0					
Terrestrial Total	25	0.14		22	12.9				
Grand Total	176	1.00		139	10.4	1.7	53	4.6	1.3

Cases of no recovery or no apparent recovery were associated with short study duration, inappropriate spill removal and clean-up techniques, lack of clean-up in experimental spills, or complex interactions with other natural or human-derived factors.

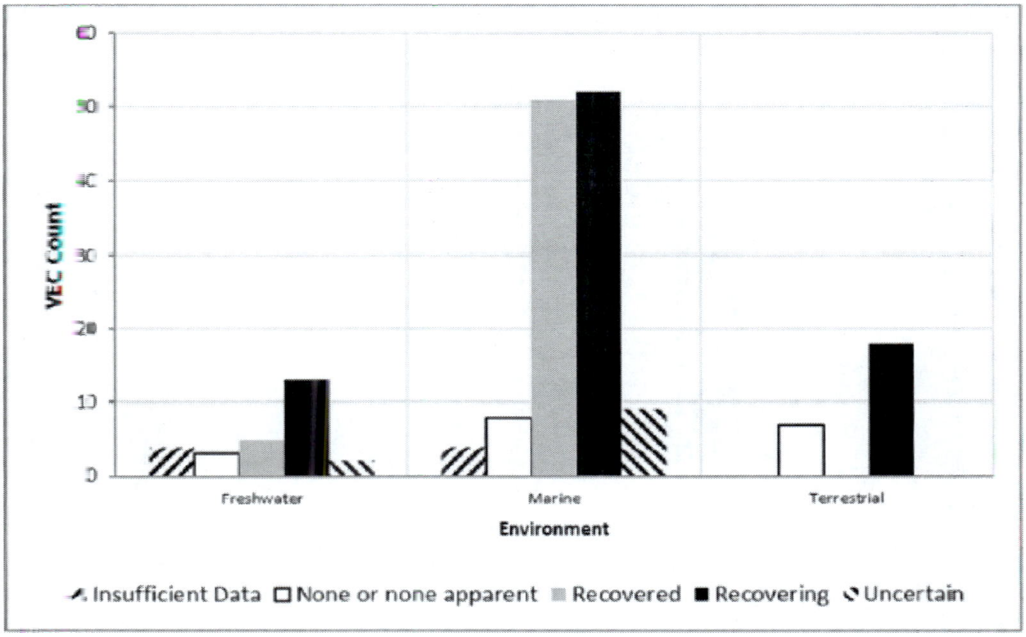

Figure 7 Recovery status by environment

How Much Time Does Recovery Take?

The environment, the VEC, and other factors influence recovery times (Table 3, Figure 8). The average times to recovered status for the biophysical VECs examined here (n=176) were 1.7 and 4.9 years for freshwater and marine environments, respectively.

The time for the VECs (n=21) in the experimental spills to achieve a recovering status ranged from 1 to 25 years with a mean of 8.9 years (Appendix). Over all environments and platforms (n=53), average recovery time was 4.6 years.

Recovery times for terrestrial VECs were complicated by the fact that seven of the terrestrial spills were experimental spills in which the oil contamination was deliberately not cleaned up. Terrestrial VECs that achieved recovering status appeared to do so in about 2 years, but some terrestrial VECs were still recovering or showed no recovery after 20 years, especially among those from experimental spills.

Fifteen of the terrestrial VECs were from Arctic and Subarctic Zones where cold temperatures decrease biodegradation rates and short growing seasons decrease annual growth rates.

Over all environments, time to recovery varied little with oil type (Figure 8b) or platform except for one instance from a refinery spill (Figure 9). Recovery time was shortest (less than a year) for spill volumes less than 1,000 metric tons (Figure 10). Recovery time was greatest (about 4 years) for spill volumes between 10,000 and 100,000 metric tons.

Recovery times varied the most across VEC categories (Figure 11). Marine mammals had the longest recovery time (about 16 years) followed by vegetation (about 10 years) and birds (about 9 years). Fish, macroinvertebrates, shorelines, and sediments had recovery times between 3 and 4 years. Algae and water quality had the shortest average recovery times of approximately 1 year.

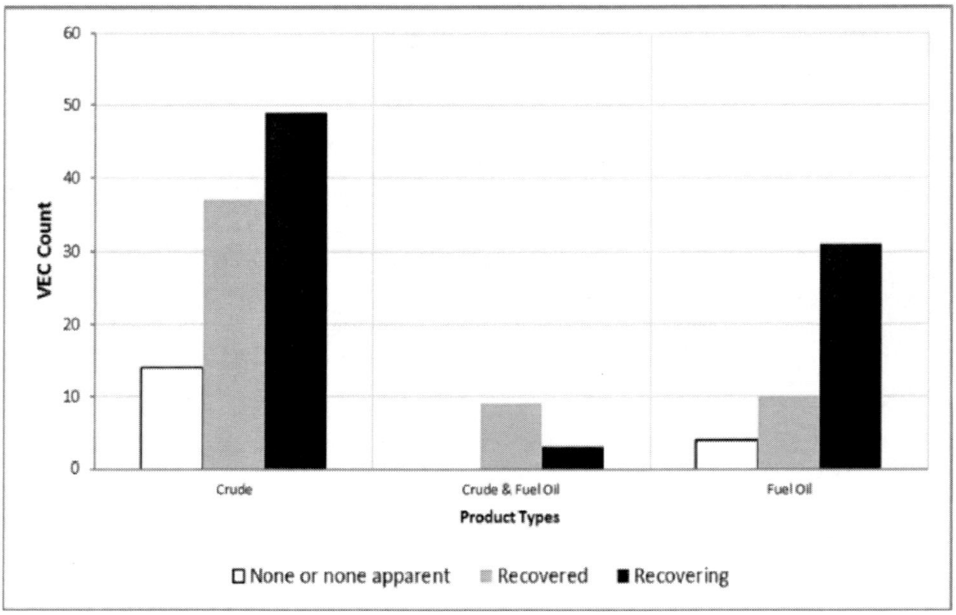

a

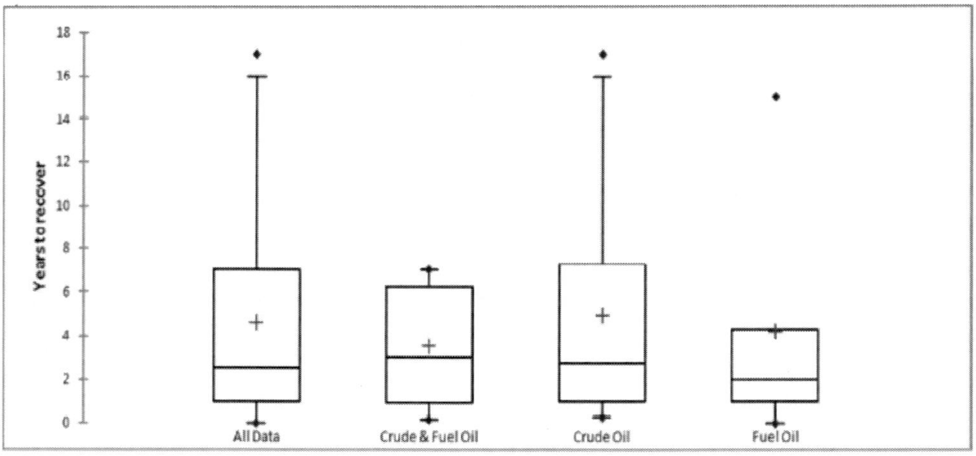

b

Figure 8. a) VEC counts by recovery status and product type, b) Recovery time by product type. Plus signs represent the mean, the boxes represent the 25th, 50th, and 75th percentiles. Minimums and maximums are represented by dots above the whiskers. This convention applies to all box plots presented in subsequent figures.

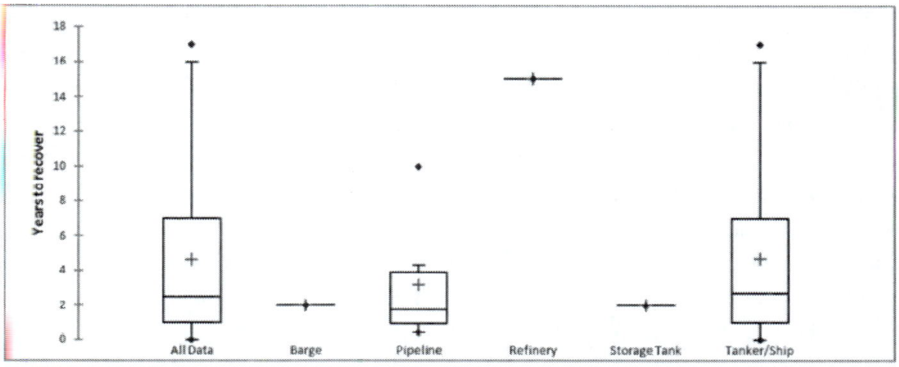

Figure 9. Recovery time by spill source.

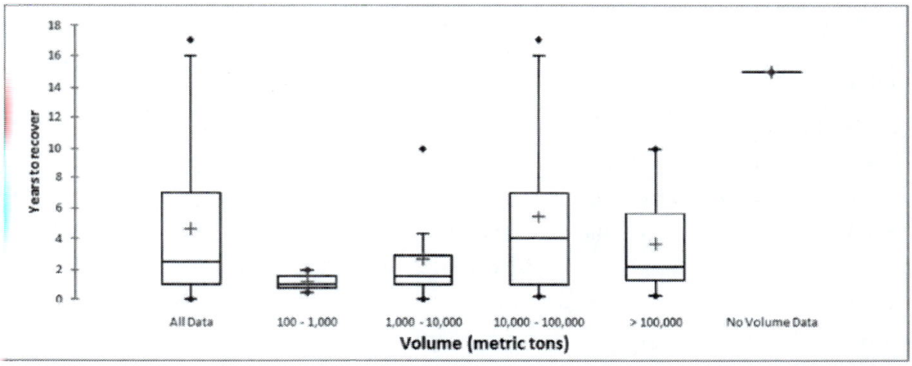

Figure 10. Recovery time by spill volume.

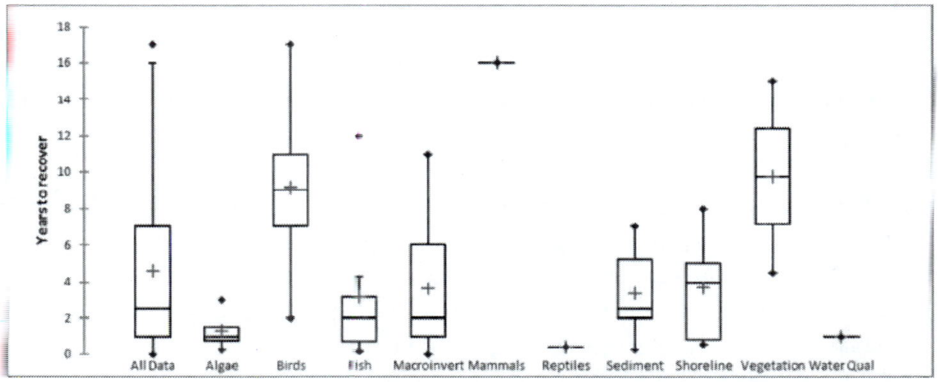

Figure 11. Recovery time by VEC.

What Processes Influence Recovery?

Any consideration of the processes in recovery from oil spills must point out the crucial role played by clean-up in enabling and accelerating recovery. One lesson learned in case studies of the recovery of estuarine and coastal ecosystems from environmental degradation is that the stressor or disturbance agent needs to be reduced to some extent before recovery can progress (Borja et al. 2010). A corollary is that both human interventions such as oil spill clean-up and natural processes such as biodegradation do not need to fully eliminate all contamination before recovery can progress. Recovery from marine spills is influenced by the degree to which the system is sheltered from physical oceanographic processes (Clark 1982, Teal and Howarth 1984). At one extreme are open marine waters that recover rapidly in weeks or months, and at the other are sheltered, soft-sediment marshes that recover slowly over two or more decades (Teal and Howarth 1984). In the middle are headlands and exposed rocky shores that take 1 to 4 years to recover (Baker 1999). Similarly, oil will persist longer in freshwater habitats, such as marshes, that are sheltered from physical processes or which sequester oil. Because of the longer recovery times for sheltered systems, modern spill response gives high priority to preventing oil from entering marshes and other similar systems.

Recovery is also influenced by several factors related to the life history of the organisms involved. First, recovery proceeds more rapidly when there is an abundant supply of propagules close to the affected area. Pelagic larvae in marine and freshwater environments enable recruitment from adjacent non-oiled areas. Kubach et al. (2010) found that recovery in a riverine environment occurred rapidly near the spill point but required longer further downstream. The differing recovery rates may be partially explained by the proximity of propagules from nearby unaffected areas. Second, the season when the spill occurred can influence the nature of the effect and the time course of recovery. For example, when a spill occurs during or after the season for colonization of invertebrates on an oiled rocky intertidal substrate, recovery may be delayed for another cycle (i.e., the next period of colonization). Third, a long life span generally leads to a long recovery time. Long-lived birds and mammals have longer recovery times than fish (Figure 11). Recovery can be longer when population growth by local reproduction rather than by immigration from other areas is the primary process in recovery.

What Factors Impede Recovery and What Can Be Done When Recovery Lags?

Several factors can slow recovery. First, if substantial levels of oil persist, recovery times may be lengthened. Clean-up is undertaken to remove and lessen the persistence of hydrocarbons. When clean-up reduces the level of contamination, a point may be reached where the environmental cost of further clean-up exceeds the benefit. In such cases, a Net Environmental Benefit Assessment (NEBA) is now often applied.

Second, inappropriate clean-up techniques can substantially lengthen recovery times. For example, Baker (1999) found that rocky shores typically recover between 1 to 4 years. However, two exceptions, *Esso Bernicia* and *Torrey Canyon,* required at least 10 years (Baker 1999). These longer recovery times were associated with bull-dozing of the shore in the *Esso Bernicia* and the use of first generation dispersants in the *Torrey Canyon*. Another example derives from the *Amoco Cadiz*, in which removal of oiled sediment from salt marshes with heavy earth-moving equipment lowered the level of the marshes and changed the patterns of sediment deposition and conditions for the growth of marsh plants. Ideally, modern spill response includes procedures to carefully select the most appropriate treatment for the oil type, level of contamination, and the nature of the shoreline.

Although the experimental spills make clear that lack of clean-up will substantially lengthen recovery times, the experimental outcomes also make clear that the most protracted recoveries in terrestrial environments appear to be where oil reaches groundwater, where cold temperatures decrease biodegradation rates in soil, or when short thaw seasons limit vegetation growth.

For such cases, recovery may benefit from active human intervention beyond routine clean-up operations. The state of the art in bioremediation and phytoremediation continually improves and holds promise for spills where natural environmental processes for oil degradation are inhibited.

How Does Recovery from Oil Spills Compare to Recovery from Other Disturbances?

Other reviews of ecosystem recovery found frequencies of recovery similar to those presented in this chapter (79% recovered and recovering overall). Jones and Schmitz (2009) examined 240 studies concerning a broad range of human disturbances and events for evidence of ecosystem recovery. For all types of disturbances, 34% of the indicator variables had recovered and 38% were recovering. For oil spills, about 60% of the indicator variables examined showed recovery. For disturbances, such as overfishing, logging, mining, eutrophication, and invasive species, recovery frequencies of indicator variables were less than 50%.

Borja et al. (2010) reviewed 53 degraded environments to compare the time course of recoveries for different types of stressors. Oils spills have moderate recovery rates. About a half of the stressor categories show recovery in less than five years, some in less than two years. Oil spills showed recovery in two to ten years. More extensive or intensive disturbances, such as disposal of sewage sludge and mine tailings, land reclamation, and long-term wastewater discharge, have shown recovery times over ten to twenty years. A widespread observation among investigators studying the effects oil spills and other disturbances is that the longer the time taken for a VEC to recover, the greater the risk that some natural event or other man-made factor will influence the course of recovery or active restoration (Parker and Wiens 2005; Jackson and Hobbs 2009; Borja 2010; Pearson et al. 2012, 2013).

What Implications Do the Survey Results Have for Oil Spill Preparedness, Prevention and Response?

Modern spill response has a much different character than in the times of the AMOCO CADIZ (1976) and the later EVOS (1989). Since the EVOS and the passage of Oil Pollution Act of 1990 in the United States, modern spill response has evolved to be able to produce more effective clean-up and meet other objectives. Built on lessons learned, the improved aspects include but are not limited to the following:

- Better state of preparedness;
- Response teams that are trained and then kept ready by regular training and drills;
- Contingency plans and maps identify and locate important resources and indicate both appropriate and inappropriate treatment options;
- Equipment deployed on a distributed basis, often on the basis of potential risk;
- Better equipment and procedures;
- Improvements in remote sensing and other instrumentation and more accurate oil spill trajectory models enable more effective allocation of response personnel and equipment;
- Incident Command Structure - a major improvement in organization and operations during a spill.

A response to a spill today would be better managed and better informed in the application of improved equipment and procedures by trained personnel than for these earlier spill events.

What Implications Do the Survey Results Have for Defining and Measuring Recovery

Any definition of recovery needs to take into account the dynamic, ever-changing and highly variable character of natural systems. Jackson and Hobbs (2009) offer a perspective:

> "First, environmental and ecological changes are normal; perhaps the most natural feature of the world in which we find ourselves is its continual flux."

Oil spills occur within these continuously changing environments. Further, it is within this continual change leading to alternate ecosystem states along multiple possible pathways that recovery of the environment from oil spills runs its course.

Recovery status and recovery time depend on how recovery is defined; this relationship has been observed in reviews of oil spills since the early 1980s through the present day (Teal and Howarth 1984; Kingston 2002; Jones and Schmitz 2009; Borja et al. 2010). These cited reviews all identify problems with defining recovery as a return to historical conditions. Again, Jackson and Hobbs (2009) offer a perspective:

> "In the long run, no inherent [particular] natural ecosystem or landscape configuration exists for any region."

Defining recovery as a return to a functioning state that provides valuable ecological goods and services is becoming increasingly seen as more appropriate.

Definitions of Recovery

Definitions of ecological (biophysical) recovery occur in both regulatory and scientific contexts. The core element in most definitions in both contexts is a return of an ecosystem or a VEC to some desirable system state following a disturbance. The details of the definitions have changed substantially over the decades as understanding of the dynamics of natural systems has increased. Generally, the changes involve the state to which the system is to return. Jackson and Hobbs (2009) offer more detail on how these changes have been influenced by advances in ecological knowledge derived from the success and failures of ecological restoration.

The ways in which recovery has been defined include return to:

- a natural equilibrium
- previous historical conditions
- pristine conditions
- some notional baseline state
- natural variability derived from comparisons of affected and reference areas
- those conditions that "might have been but for the spill"
- a functional ecosystem state that provides valuable ecological goods and services.

Ecologists have moved away from using such recovery end states such as return to equilibrium because of the increasing recognition of the natural dynamics that lead to ecosystem change, rather than any equilibrium (Landis 2007; Wu and Loucks 1995; Kapustka and Landis 1998; Jackson and Hobbs 2009). For many, any equilibrium in the ecological realm now appears to be only remotely possible, if at all. Return to a previous historical state has also proved problematic and impractical because our understanding of past historical system states is inadequate (Jackson and Hobbs 2009). Also, achieving restoration to previous historical states has frequently failed because other historical natural and man-made events have produced fundamental changes in the system.

Similarly, the notion of a return to pristine conditions is problematic and impractical because pristine conditions are too often an assumption rather than a reality. For example, the Prince William Sound (PWS) ecosystem, despite prevailing beliefs, was not pristine at the time of the EVOS (Wooley 2002). Human activities such as commercial harvesting of fur, fish, and timber, as well as mining and the introduction of non-indigenous species, had already produced extensive ecosystem transformations in the PWS ecosystems. Also, just prior to the EVOS, substantial declines were evident in harbor seal and Steller sea lion populations (Braham et al. 1980; Calkins et al. 1994; Frost et al. 1994), and hatcheries were adding over 500 million juvenile pink salmon to PWS each year (Pearson et al. 2012).

Return to baseline often involves a comparison of post-spill conditions with "baseline" conditions measured usually for a single VEC species or species guild during some pre-project phase of development. Where such baselines have been based on only a year or two of study, they have proven to be inadequate, because basic system dynamics can require years to fully assess. Also, the baseline sampling often proves to be in the "wrong" place to provide "before" data.

Defining recovery as a return to the natural variability of the resource was intended to address some of the shortcomings of the comparison-to-baseline approach. This definition assumes that recovery occurs when the resource in the affected area is "tracking" (i.e., changing in parallel with) the changes in the resource in an unaffected reference area. The subsection below on measuring recovery gives more details on the ways in which this definition can be assessed.

Recently, recovery has been defined as a return to the conditions that would have prevailed had the oil spill not occurred. This definition recognizes the need to account for natural variability and for the influence of natural and man-made factors other than the spill.

Jackson and Hobbs (2009) also argue that the goal of recovery and of active restoration of damaged environments should be to recover or restore the ecosystem to a functional state that provides valuable ecological goods and services. This perspective is also emerging in the regulatory context in North America.

Measuring Recovery

Several reviews compare study designs and ways to measure recovery following oil spills (Wiens and Parker 1995; Parker and Wiens 2005; Wiens 2013), including:

- comparison to baseline
- single-year comparison of affected versus reference area
- multi-year comparisons of affected versus reference areas
- Before-After-Control-Impact (BACI) studies
- weight of evidence

Parker and Wiens (2005) conclude that there is no one "best" way to assess recovery and urge future oil spill investigators to discern how the ecosystem to be studied functions, and determine which set of assumptions prevail for that ecosystem before selecting an appropriate study approach. However, the BACI approach appears to be increasingly favored for assessing oil spill recovery (Parker and Wiens 2005). Where "before" is lacking, the investigator has to fall back a Control – Impact design and wrestle with locating an appropriate control area to compare with the affected area (Parker et al. 2013, Pearson 2013).

One example from EVOS provides some vivid illustrations of how recovery definition influence study outcomes. Landis (2007) compares the different conclusions about recovery from the EVOS in two prominent papers (Peterson et al. 2003, Harwell and Gentile 2006) and seeks to understand why the conclusions differ so much. Landis finds in part that the differences between the two papers derive from "the infusion of social values or policy goals into each." However, the more important source of the differences for Landis (2007) is how the two papers approach the definition of recovery and the role of indirect effects and effect cascades. Indeed, Peterson et al. (2003) do not offer an explicit definition of recovery and appear to imply that recovery is the return to some undefined previous state. In contrast, Harwell and Gentile (2006) do offer a recovery definition – return to an acceptable range of variability in the resource of interest – but one that includes a policy statement ("acceptable") and one for which the outcome rests on a judgment of acceptability rather than on a scientific operation such as a statistical comparison of the attributes of two areas or populations. Landis (2007) recommends that researchers achieve and communicate clearly expressed operational definitions and goal statements in scientific assessments of effects and recovery.

Another aspect of measuring recovery emerges from the assessment here. The review of the studies often observed that when resources were categorized as "recovering," no additional studies could be found verifying "recovered." One plausible explanation for this observation is that the participating parties were sufficiently comfortable with the likelihood of recovery that funding was re-allocated and full recovery was never actually documented.

CONCLUSION

The scientific literature on recovery is uneven, with marine oils spills and certain VECs receiving the bulk of the studies. We suggest that this condition derives from the fact that tanker and ship accidents spill substantially larger volumes of oil that lead to effects over larger areas and for more VECs. Also, the freshwater studies show substantially less time to recovery than marine spills. We further suggest that the literature does not provide a full understanding of recovery following terrestrial spills. For some presumably large but unknown number of terrestrial spills, oiled soil and vegetation is removed for treatment elsewhere and the areas replanted. In such cases, it appears that recovery is simply not monitored. Clearly, improved response and remediation would benefit from greater study of recovery from oils spills in freshwater and terrestrial systems.

Although adverse effects of oil spills on biophysical environments cannot be denied, the scientific literature is clear that ecosystems and their components do recover. Cases of no or no apparent recovery were associated with short study duration, inappropriate spill response procedures, experimental spills, and complex interactions with other natural or human-derived factors. Recovery times proved to vary with the environment, the VEC, and other factors. The literature is also clear that although recovery can occur in some circumstances with little or no clean-up, appropriate clean-up does enable and can accelerate recovery.

Our overall observation that biophysical environments do recover should not be taken to indicate spill response can be foregone. Spill response is most definitely needed and the capacity for such response needs to be established beforehand and kept in a state of readiness. Such effort will lead to timely and appropriate response that will enable and accelerate recovery.

ACKNOWLEDGMENTS

We gratefully acknowledge the support of the Northern Gateway Project for the previous efforts leading to this chapter. Preparation of this chapter was supported by internal funding from Stantec Consulting Services, Inc. Joey Seamands produced the map presented as Figure 1. Michael Johnson, Joey Seamands, Jake Riley, Nathan Henderson, Lindsay Conley, and Liz Bockstiegel were instrumental in the construction of the early versions of the database.

REFERENCES

American Petroleum Institute. (1999). Fate and Environmental Effects of Oil Spills in Freshwater Environments. In D. Stalfort (Ed.), *API Publication Number 4675* (pp. 158).

Baker, J. M. (1999). Ecological Effectiveness of Oil Spill Countermeasures: how clean is clean? *Pure and Applied Chemistry, 71*(1), 135-151.

Borja, Á., Dauer, D., Elliott, M., & Simenstad, C. (2010). Medium- and Long-term Recovery of Estuarine and Coastal Ecosystems: Patterns, Rates and Restoration Effectiveness. *Estuaries and Coasts, 33*(6), 1249-1260. doi: 10.1007/s12237-010-9347-5.

Braham, H., Everitatn, D. D., & Rugh, J. (1980). Northern Sea Lion Population Decline in the Eastern Aleutian Islands. *Journal of Wildlife Management, 44*(1), 25-33.

Calkins, D. G., Becker, E. F., Spraker, T. R., & Loughlin, T. R. (1994). *Assessment of Injury to Steller Sea Lions in Prince William Sound and the Gulf of Alaska.* (GC1552.P75 E99 MMS-04). Retrieved from http://www.evostc.state.ak.us/Files.cfm?doc=/Store/FinalReports/1992-MM04-Final.pdf.

Clark, R. B. (1982). The Long-Term Effects of Oil Pollution on Marine Populations, Communities, and Ecosystems. *Philosophical Transactions of the Royal Society of London. Series B, Biological Sciences, 297*(1087), 185-192.

Collins, C. M., Racine, C. H., & Walsh, M. E. (1994). The physical, chemical, and biological effects of crude oil spills after 15 years on a black spruce forest, Interior Alaska. *ARCTIC, 47*(2), 164-175.

Fall, J. A., Walker, R. J., Stanek, R. T., Simeone, W. E., Hutchinson-Scarborough, L., Coiley-Kenner, P., . . . Koster, D. (2006). *Update of the Status of Subsistence Uses in Exxon Valdez Oil Spill Area Communities, 2003.* (312). Juneau, AK: James A. Fall.

Frost, K. J., Lowry, L. F., Sinclair, E. F., Hoef, J. V., & MaAllister, D. C. (1994). Impacts on distribution, abundance, and productivity of harbor seals. *Marine Mammals and the Exxon Valdez* (pp. 97-118). San Diego, California: Academic Press.

Harwell, M. A., & Gentile, J. H. (2006). Ecological Significance of Residual Exposures and Effects from the Exxon Valdez Oil Spill. *Integrated Environmental Assessment & Management, 2*(3), 204-246.

Jackson, S. T., & Hobbs, R. J. (2009). Ecological Restoration in the Light of Ecological History. *Science, 325*(5940), 567-569.

Jones, H. P., & Schmitz, O. J. (2009). Rapid recovery of damaged ecosystems. *PLoS ONE, 4*(5), e5653. doi: 10.1371/journal.pone.0005653.

Kapustka, L. A., & Landis, W. G. (1998). Ecology: The Science Versus the Myth. *Human and Ecological Risk Assessment: An International Journal, 4*(4), 829-838. doi: 10.1080/10807039891284820.

Kingston, P. F. (2002). Long-term Environmental Impact of Oil Spills. *Spill Science & Technology Bulletin, 7*(1-2), 53-61.

Kubach, K. M., Scott, M. C., & Bulak, J. S. (2011). Recovery of a Temperate Riverine Fish Assemblage from a Major Diesel Oil spill. *Freshwater Biology, 56*, 503-518. doi: 10.1111/j.1365-2427.2010.02517.x.

Landis, W. G. (2007). The Exxon Valdez Oil Spill Revisited and the Dangers of Normative Science. *Integrated Environmental Assessment & Management, 3*(3), 439-441.

National Research Council. (1975). *Oil in the Marine Environment.* Washington, D.C.: National Academy Press.

National Research Council. (1985). *Oil in the Sea: Inputs, Fates, and Effects Steering Committee for the Petroleum in the Marine Environment Update,* (pp. 601). Retrieved from http://www.nap.edu/catalog/314.html.

National Research Council. (1999). *Spills on Nonfloating Oils: Risk and Response.* Washington, D.C.: National Academy Press.

National Research Council. (2003). *Oil in the Sea III: Inputs, Fates, and Effects Committee on Marine Transportation of Heavy Oils,* (pp. 278). Retrieved from http://www.nap.edu/catalog/10388.html.

Parker, K. R., & Wiens, J. A. (2005). Assessing recovery following environmental accidents: environmental variation, ecological assumptions, and strategies. *Ecological Applications, 15*(6), 2037-2051.

Parker, K. R., Wiens, J. A., Day, R. H., & Murphy, S. M. (2013). Assessing Effects and Recovery from Environmental Accidents. In J. A. Wiens (Ed.), *Oil in the Environment: Legacies and Lessons of the Exxon Valdez Oil Spill* (pp. 220-240). Cambridge, U.K., 2013: Cambridge University Press.

Pearson, W., Deriso, R., Elston, R., Hook, S., Parker, K., & Anderson, J. (2012). Hypotheses concerning the decline and poor recovery of Pacific herring in Prince William Sound, Alaska. *Reviews in Fish Biology & Fisheries, 22*(1), 95-135. doi: 10.1007/s11160-011-9225-7.

Pearson, W. H., Elston, R. A., Humphrey, K., & Deriso, R. B. (2013). Pacific herring. In J. A. Wiens (Ed.), *Oil in the Environment: Legacies and Lessons of the Exxon Valdez Oil Spill* (pp. 292-317). Cambridge, U.K.: Cambridge University Press.

Peterson, C. H., Rice, S. D., Short, J. W., Esler, D., Bodkin, J. L., Ballachey, B. E., & Irons, D. B. (2003). Long-Term Ecosystem Response to the Exxon Valdez Oil Spill. *Science, 302*(5653), 2082-2086.

Prince, R. C., Garrett, R. M., Bare, R. E., Grossman, M. J., Townsend, T., Suflita, J. M., . . . Lessard, R. R. (2003). The Roles of Photooxidation and Biodegradation in Long-term Weathering of Crude and Heavy Fuel Oils. *Spill Science & Technology Bulletin, 8*(2), 145-156. doi: 10.1016/s1353-2561(03)00017-3.

Rice, S. D., Spies, R. B., Wolfe, D. A., & Wright, B. A. (Eds.). (1996). *Proceedings of the Exxon Valdez Oil Spill Symposium*. Bethesda, MD: American Fisheries Society.

Stantec Consulting Ltd., AMEC Earth and Environment, Chumis Cultural Resource Services, & Coastal Assessment Liaison and Monitoring. (2012). Reply Evidence: Recovery of the Biophysical and Human Environments from Oil Spills, Enbridge Northern Gateway Project *Prepared for Northern Gateway Pipelines Limited Partnership* (pp. 264).

Teal, J. M., & Howarth, R. W. (1984). Oil spill studies: A review of ecological effects. *Environ Manage, 8*(1), 27-43.

Wells, P. G., Butler, J. N., & Hughes, J. S. (1995). *Exxon Valdez Oil Spill: Fate and Effects in Alaskan Waters* (P. G. Wells, J. S. Butler & J. S. Hughes Eds. Vol. 1219). Philadelphia, PA: Astm STP Intl (February 1996).

Whittle, K. J., Hardy, R., Mackie, P. R., & McGill, A. S. (1982). A quantitative assessment of the sources and fates of petroleum compounds in the marine environment. In R. B. Clark (Ed.). *The Long-Term Effects of Oil Pollution on Marine Populations, Communities, and Ecosystems*. London: The Royal Society.

Wiens, J. A. (2013). *Oil in the Environment: Legacies and Lessons of the Exxon Valdez Oil Spill*. Cambridge, U.K.: Cambridge University Press.

Wiens, J. A., & Parker, K. R. (1995). Analyzing the effects of accidental environmental impacts: Approaches and assumptions. *Ecological Applications, 5*(4), 1069.

Wooley, C. (2002). The myth of the ''pristine environment'': Past human impacts in Prince William Sound and the Northern Gulf of Alaska. *Spill Science Technical Bulletin, 7*, 89-104.

Wu, J., & Loucks, O. (1995). From the balance of nature to hierarchical patch dynamics: A paradigm shift in ecology. *Quarterly Review of Biology, 70*, 439-466.

APPENDIX

Oil Spill Name	Year	Product Type (general)	Volume (metric tons) total	Platform	Environments	VEC *	Recovery Status	Years to recover	Study Duration	References
AMOCO CADIZ	1978	Light crude	240,000	Tanker/Ship	Marine	ALG	Recovered	0.25	UNK	Spooner 1978
AMOCO CADIZ	1978	Light crude	240,000	Tanker/Ship	Marine	FSH	Recovered	0.5	UNK	Seymour and Geyer 1992
AMOCO CADIZ	1978	Light crude	240,000	Tanker/Ship	Marine	MCR	Recovered	1	UNK	Seymour and Geyer 1992
AMOCO CADIZ	1978	Light crude	240,000	Tanker/Ship	Marine	MCR	Recovered	2	6	Seymour and Geyer 1992
AMOCO CADIZ	1978	Light crude	240,000	Tanker/Ship	Marine	MCR	Recovered	2	18	Kingston 2002 Dauvin 1998
AMOCO CADIZ	1978	Light crude	240,000	Tanker/Ship	Marine	MCR	Recovered	2.5	UNK	Seymour and Geyer 1992
AMOCO CADIZ	1978	Light crude	240,000	Tanker/Ship	Marine	MCR	Recovered	6	18	Kingston 2002
AMOCO CADIZ	1978	Light crude	240,000	Tanker/Ship	Marine	MCR	Recovered	10	19	Kingston 2002, Dauvin 1998, Dauvin 2000
AMOCO CADIZ	1978	Light crude	240,000	Tanker/Ship	Marine	MCR	Recovering	UNK	UNK	Glemarec and Hussenot 1982
AMOCO CADIZ	1978	Light crude	240,000	Tanker/Ship	Marine	SHR	Recovered	8	18	Kingston 2002
AMOCO CADIZ	1978	Light crude	240,000	Tanker/Ship	Marine	VEG	Recovered	4.5	8	Baca et al. 1987
ARROW	1970	No. 6 Fuel Oil	7,980	Tanker/Ship	Marine	ALG	Recovering	N/A	7	Thomas 1978
ARROW	1970	No. 6 Fuel Oil	7,980	Tanker/Ship	Marine	FSH	Recovering	N/A	30	Lee et al. 2003
ARROW	1970	No. 6 Fuel Oil	7,980	Tanker/Ship	Marine	MCR	Recovering	N/A	6	Gilfillan 1978

Oil Spill Name	Year	Product Type (general)	Volume (metric tons) total	Platform	Environments	VEC *	Recovery Status	Years to recover	Study Duration	References
ARROW	1970	No. 6 Fuel Oil	7,980	Tanker/Ship	Marine	MCR	Recovering	N/A	30	Lee et al. 2003
ARROW	1970	No. 6 Fuel Oil	7,980	Tanker/Ship	Marine	MCR	Recovered	UNK	30	Lee et al. 2003
ARROW	1970	No. 6 Fuel Oil	7,980	Tanker/Ship	Marine	SED	Recovering	N/A	30	Lee et al. 2003
ARROW	1970	No. 6 Fuel Oil	7,980	Tanker/Ship	Marine	SHR	Recovering	N/A	7	Thomas 1978
ARROW	1970	No. 6 Fuel Oil	7,980	Tanker/Ship	Marine	SHR	Recovering	N/A	27	Lee et al 1999
ARROW	1970	No. 6 Fuel Oil	7,980	Tanker/Ship	Marine	SHR	Recovering	N/A	33	Owens et al. 1978, Owens et al. 2008, Prince et al. 2003
ARROW	1970	No. 6 Fuel Oil	7,980	Tanker/Ship	Marine	SHR	None	N/A	UNK	Kingston 2002
Arthur Kill	1990	No. 6 Fuel Oil & No. 2 Fuel Oil	4,216	Pipeline	Marine	BRD	Recovered	2.5	UNK	Parsons 1996
Arthur Kill	1990	No. 6 Fuel Oil & No. 2 Fuel Oil	4,216	Pipeline	Marine	BRD	Recovered	10	10	Maccarone and Brzorad 2000
Asher Creek	1979	Crude	1,000	Pipeline	Freshwater	MCR	Recovered	0.92	1.46	Crunkilton and Duchrow 1990; Seymour and Geyer 1992
Ashland Spill	1988	No. 2 Fuel Oil	2,234	Storage Tank	Freshwater	BRD	Recovered	2	2	University of Pittsburgh 1990
Ashland Spill	1988	No. 2 Fuel Oil	2,234	Storage Tank	Freshwater	SED	Recovering	N/A	2	University of Pittsburgh 1990
Baffin Island (BIOS)	1981	Crude	15	Experimental	Marine	SHR	Recovering	N/A	20	Owens et al. 1987, Owens et al. 2008, Humphrey et al. 1991, Prince et al. 2002, Prince et al. 2003
BAHIA PARAISO	1989	No. 2 Fuel Oil	500	Tanker/Ship	Marine	MCR	Recovering	N/A	1	Seymour and Geyer 1992
Bemidji	1979	Crude	1459	Pipeline	Terrestrial	GWR	Recovering	N/A	30	Revesz et al. 1995, Essaid et al. 2011
Bemidji	1979	Crude	1459	Pipeline	Terrestrial	GWR	Recovering	N/A	33	USGS 2012

Appendix (Continued)

Oil Spill Name	Year	Product Type (general)	Volume (metric tons) total	Platform	Environments	VEC *	Recovery Status	Years to recover	Study Duration	References
BRAER	1993	Crude & No. 6 Fuel Oil	85,475	Tanker/Ship	Marine	FSH	Recovered	0.17	UNK	Kingston 1999
BRAER	1993	Crude & No. 6 Fuel Oil	85,475	Tanker/Ship	Marine	FSH	Recovering	N/A	UNK	Stagg et al. 1998
BRAER	1993	Crude & No. 6 Fuel Oil	85,475	Tanker/Ship	Marine	MCR	Recovered	1	UNK	Kingston 1999
BRAER	1993	Crude & No. 6 Fuel Oil	85,475	Tanker/Ship	Marine	MCR	Recovered	7	UNK	Webster et al. 2005
BRAER	1993	Crude & No. 6 Fuel Oil	85,475	Tanker/Ship	Marine	MCR	Recovering	N/A	UNK	Kingston 1999
BRAER	1993	Crude & No. 6 Fuel Oil	85,475	Tanker/Ship	Marine	SED	Recovered	7	UNK	Webster et al. 2005
BRAER	1993	Crude & No. 6 Fuel Oil	85,475	Tanker/Ship	Marine	SHR	Recovered	0.83	UNK	Kingston 1999
BRAER	1993	Crude & No. 6 Fuel Oil	85,475	Tanker/Ship	Marine	SHR	Recovered	UNK	UNK	Kingston 2002
Caribou-Poker	1976	Crude	7	Experimental	Terrestrial	MCB	None	N/A	0.42	Sparrow et al. 1978
Caribou-Poker	1976	Crude	7	Experimental	Terrestrial	MCB	None	N/A	19	Lindstrom et al. 1999
Caribou-Poker	1976	Crude	7	Experimental	Terrestrial	SOI	Recovering	N/A	10	Sparrow and Sparrow 1988
Caribou-Poker	1976	Crude	7	Experimental	Terrestrial	SOI	None	N/A	15	Collins et al. 1994, Prince et al. 2003
Caribou-Poker	1976	Crude	7	Experimental	Terrestrial	SOI	Recovering	N/A	25	Braddock et al. 2003
Caribou-Poker	1976	Crude	7	Experimental	Terrestrial	SOI	Recovering	N/A	25	Prince et al. 2003
Caribou-Poker	1976	Crude	7	Experimental	Terrestrial	SOI	None	N/A	UNK	Fitzgerald 2007-2008
Caribou-Poker	1976	Crude	7	Experimental	Terrestrial	VEG	None	N/A	2	Jenkins et al. 1978
Caribou-Poker	1976	Crude	7	Experimental	Terrestrial	VEG	None	N/A	15	Collins et al. 1994
Caribou-Poker	1976	Crude	7	Experimental	Terrestrial	VEG	Recovering	N/A	20	Racine 1994

Oil Spill Name	Year	Product Type (general)	Volume (metric tons) total	Platform	Environments	VEC *	Recovery Status	Years to recover	Study Duration	References
Cayuga Inlet	1997	No. 2 Fuel Oil	23	Rail	Freshwater	MCR	Recovering	N/A	1.25	Lytle and Peckarsky 2001
Chalk Point	2000	No. 6 Fuel Oil & No. 2 Fuel Oil	447	Pipeline	Marine	MCR	Insufficient Data	N/A	0.5	Versar 2001
Chalk Point	2000	No. 6 Fuel Oil & No. 2 Fuel Oil	447	Pipeline	Marine	MCR	Recovering	N/A	7	Michel et al. 2009
Chalk Point	2000	No. 6 Fuel Oil & No. 2 Fuel Oil	447	Pipeline	Marine	SED	Recovering	N/A	7	Michel et al. 2009
Chalk Point	2000	No. 6 Fuel Oil & No. 2 Fuel Oil	447	Pipeline	Marine	VEG	Uncertain	N/A	7	Michel et al. 2009
Chariton River	1990	Light crude	356	Pipeline	Freshwater	MCR	Recovering	N/A	1	Poulton et al. 1998
COSCO BUSCAN	2007	No. 6 Fuel Oil	184	Tanker/Ship	Marine	FSH	Uncertain	N/A	UNK	California Department of Fish and Game 2009, California Department of Fish and Game 2010, California Department of Fish and Game 2011
Delaware Mesocosm	1994	Crude	0.1	Experimental	Marine	SHR	Insufficient Data	N/A	1.23	Venosa and Zhu 2003
East Walker River	2000	No. 6 Fuel Oil	11	Truck	Freshwater	SED	Recovering	N/A	0.41	Higgins et al. 2002, Hampton et al. 2002
East Walker River	2000	No. 6 Fuel Oil	11	Truck	Freshwater	WQL	Recovering	N/A	0.41	Higgins et al. 2002
ERIKA	1999	No. 6 Fuel Oil	28,000	Tanker/Ship	Marine	SHR	Recovering	N/A	0.3	Prince et al. 2003
ESSO BERNICIA	1978	No. 6 Fuel Oil	1,174	Tanker/Ship	Marine	ALG	Recovered	1	16	Moore et al. 1995
ESSO BERNICIA	1978	No. 6 Fuel Oil	1,174	Tanker/Ship	Marine	MCR	Recovered	1	16	Moore et al. 1995

Appendix (Continued)

Oil Spill Name	Year	Product Type (general)	Volume (metric tons) total	Platform	Environments	VEC *	Recovery Status	Years to recover	Study Duration	References
Experimental Study, LA	1973	Light crude	ND	Experimental	Freshwater	BRD	Uncertain	N/A	0.5	Chabreck 1973
EXXON VALDEZ	1989	Crude	33,000	Tanker/Ship	Marine	ALG	Recovered	3	8	Kingston 2002, Hoff & Shigenaka 1999
EXXON VALDEZ	1989	Crude	33,000	Tanker/Ship	Marine	ALG	Recovering	N/A	7	Houghton et al. 1997
EXXON VALDEZ	1989	Crude	33,000	Tanker/Ship	Marine	BRD	Recovered	7	UNK	Bowman 1999
EXXON VALDEZ	1989	Crude	33,000	Tanker/Ship	Marine	BRD	Recovered	7	UNK	Wiens et al. 2001
EXXON VALDEZ	1989	Crude	33,000	Tanker/Ship	Marine	BRD	Recovered	9	9	Murphy and Mabee 2000
EXXON VALDEZ	1989	Crude	33,000	Tanker/Ship	Marine	BRD	Recovered	11	14	Esler and Iverson 2010
EXXON VALDEZ	1989	Crude	33,000	Tanker/Ship	Marine	BRD	Recovered	17	21	Exxon Valdez Oil Spill Trustee Council 2010
EXXON VALDEZ	1989	Crude	33,000	Tanker/Ship	Marine	BRD	Recovered	17	21	Exxon Valdez Oil Spill Trustee Council 2010
EXXON VALDEZ	1989	Crude	33,000	Tanker/Ship	Marine	BRD	None	N/A	9	Kingston 2002, Lance et al. 2001
EXXON VALDEZ	1989	Crude	33,000	Tanker/Ship	Marine	BRD	None	N/A	9	Lance et al. 2001
EXXON VALDEZ	1989	Crude	33,000	Tanker/Ship	Marine	BRD	Recovering	N/A	10	Trust et al. 2000
EXXON VALDEZ	1989	Crude	33,000	Tanker/Ship	Marine	BRD	None	N/A	20	Exxon Valdez Oil Spill Trustee Council 2010

Oil Spill Name	Year	Product Type (general)	Volume (metric tons) total	Platform	Environments	VEC *	Recovery Status	Years to recover	Study Duration	References
EXXON VALDEZ	1989	Crude	33,000	Tanker/Ship	Marine	BRD	Recovering	N/A	20	Esler et al. 2010
EXXON VALDEZ	1989	Crude	33,000	Tanker/Ship	Marine	BRD	Recovering	N/A	UNK	Exxon Valdez Oil Spill Trustee Council 2010
EXXON VALDEZ	1989	Crude	33,000	Tanker/Ship	Marine	BRD	Recovering	N/A	UNK	Exxon Valdez Oil Spill Trustee Council 2010
EXXON VALDEZ	1989	Crude	33,000	Tanker/Ship	Marine	BRD	Uncertain	N/A	UNK	Exxon Valdez Oil Spill Trustee Council 2010
EXXON VALDEZ	1989	Crude	33,000	Tanker/Ship	Marine	BRD	Recovered	UNK	UNK	Kingston 2002
EXXON VALDEZ	1989	Crude	33,000	Tanker/Ship	Marine	FSH	Recovered	1	2	Pearson et al. 1995
EXXON VALDEZ	1989	Crude	33,000	Tanker/Ship	Marine	FSH	Recovered	2	UNK	Lee and Page 1997
EXXON VALDEZ	1989	Crude	33,000	Tanker/Ship	Marine	FSH	Recovered	12	UNK	Exxon Valdez Oil Spill Trustee Council 2002
EXXON VALDEZ	1989	Crude	33,000	Tanker/Ship	Marine	FSH	Insufficient Data	N/A	3	Collier et al. 1993
EXXON VALDEZ	1989	Crude	33,000	Tanker/Ship	Marine	FSH	Recovering	N/A	10	Jewett et al. 2002
EXXON VALDEZ	1989	Crude	33,000	Tanker/Ship	Marine	MCR	Recovered	2	UNK	Hoff and Shigenaka 1999
EXXON VALDEZ	1989	Crude	33,000	Tanker/Ship	Marine	MCR	Recovered	4	8	Hoff and Shigenaka 1999
EXXON VALDEZ	1989	Crude	33,000	Tanker/Ship	Marine	MCR	Recovered	11	12	Page et al. 2005

Appendix (Continued)

Oil Spill Name	Year	Product Type (general)	Volume (metric tons) total	Platform	Environments	VEC *	Recovery Status	Years to recover	Study Duration	References
EXXON VALDEZ	1989	Crude	33,000	Tanker/Ship	Marine	MCR	Recovering	N/A	1	Brown et al. 1996
EXXON VALDEZ	1989	Crude	33,000	Tanker/Ship	Marine	MCR	Recovering	N/A	2	Driskell et al. 1993
EXXON VALDEZ	1989	Crude	33,000	Tanker/Ship	Marine	MCR	Uncertain	N/A	3	Lee and Page 1997
EXXON VALDEZ	1989	Crude	33,000	Tanker/Ship	Marine	MCR	Uncertain	N/A	6	Fukuyama et al. 2000
EXXON VALDEZ	1989	Crude	33,000	Tanker/Ship	Marine	MCR	Recovering	N/A	8	Hoff and Shigenaka 1999
EXXON VALDEZ	1989	Crude	33,000	Tanker/Ship	Marine	MML	Recovered	16	21	Exxon Valdez Oil Spill Trustee Council 2010
EXXON VALDEZ	1989	Crude	33,000	Tanker/Ship	Marine	MML	Uncertain	N/A	11	Bodkin et al. 2002
EXXON VALDEZ	1989	Crude	33,000	Tanker/Ship	Marine	MML	None	N/A	21	Exxon Valdez Oil Spill Trustee Council 2010
EXXON VALDEZ	1989	Crude	33,000	Tanker/Ship	Marine	MML	Recovering	N/A	21	Exxon Valdez Oil Spill Trustee Council 2010
EXXON VALDEZ	1989	Crude	33,000	Tanker/Ship	Marine	SED	Recovered	0.25	8	Hoff and Shigenaka 1999
EXXON VALDEZ	1989	Crude	33,000	Tanker/Ship	Marine	SED	Recovered	2	8	Hayes and Michel 1999
EXXON VALDEZ	1989	Crude	33,000	Tanker/Ship	Marine	SED	Recovered	2	UNK	Lee and Page 1997

Oil Spill Name	Year	Product Type (general)	Volume (metric tons) total	Platform	Environments	VEC *	Recovery Status	Years to recover	Study Duration	References
EXXON VALDEZ	1989	Crude	33,000	Tanker/Ship	Marine	SED	Recovered	3	8	Hayes and Michel 1999
EXXON VALDEZ	1989	Crude	33,000	Tanker/Ship	Marine	SED	Recovered	6	8	Hoff and Shigenaka 1999
EXXON VALDEZ	1989	Crude	33,000	Tanker/Ship	Marine	SED	None	N/A	8	Hayes and Michel. 1999
EXXON VALDEZ	1989	Crude	33,000	Tanker/Ship	Marine	SED	Recovering	N/A	12	Short et al. 2003
EXXON VALDEZ	1989	Crude	33,000	Tanker/Ship	Marine	SED	Recovering	N/A	18	Michel et al. 2010
EXXON VALDEZ	1989	Crude	33,000	Tanker/Ship	Marine	SHR	Recovered	0.6	UNK	Lee and Page 1997
EXXON VALDEZ	1989	Crude	33,000	Tanker/Ship	Marine	SHR	Recovered	4	8	Kingston 2002
EXXON VALDEZ	1989	Crude	33,000	Tanker/Ship	Marine	SHR	Recovering	N/A	2	Lee and Page 1997
EXXON VALDEZ	1989	Crude	33,000	Tanker/Ship	Marine	SHR	Recovering	N/A	3	Neff et al. 1995
EXXON VALDEZ	1989	Crude	33,000	Tanker/Ship	Marine	SHR	Recovering	N/A	3	Stoker et al. 1993
EXXON VALDEZ	1989	Crude	33,000	Tanker/Ship	Marine	SHR	Recovering	N/A	3.5	Michel and Hayes 1993
EXXON VALDEZ	1989	Crude	33,000	Tanker/Ship	Marine	SHR	None	N/A	4.9	Irvine et al. 1999
EXXON VALDEZ	1989	Crude	33,000	Tanker/Ship	Marine	SHR	Recovering	N/A	8	Hoff and Shigenaka 1999
EXXON VALDEZ	1989	Crude	33,000	Tanker/Ship	Marine	SHR	Recovering	N/A	16	Owens et al. 2008

Appendix (Continued)

Oil Spill Name	Year	Product Type (general)	Volume (metric tons) total	Platform	Environments	VEC *	Recovery Status	Years to recover	Study Duration	References
EXXON VALDEZ	1989	Crude	33,000	Tanker/Ship	Marine	VEG	Uncertain	N/A	1	Dean et al. 1998
EXXON VALDEZ	1989	Crude	33,000	Tanker/Ship	Marine	WQL	Recovered	1	3	Neff and Burns 1996
EXXON VALDEZ	1989	Crude	33,000	Tanker/Ship	Marine	WQL	Recovered	1	4	Boehm et al. 2007.
Fidalgo Bay	1991	Crude	670	Refinery	Marine	SED	Recovering	N/A	1.3	Hoff et al. 1993
Fidalgo Bay	1991	Crude	670	Refinery	Marine	VEG	Recovering	N/A	1.3	Hoff et al. 1993
FLORIDA	1969	No. 2 Fuel Oil	557	Barge	Marine	FSH	Recovering	N/A	20	Teal et al. 1992
FLORIDA	1969	No. 2 Fuel Oil	557	Barge	Marine	MCR	Recovering	N/A	5	Michael et al. 1975, Sanders et al. 1980
FLORIDA	1969	No. 2 Fuel Oil	557	Barge	Marine	MCR	Recovering	N/A	7	Krebs and Burns 1977
FLORIDA	1969	No. 2 Fuel Oil	557	Barge	Marine	MCR	Recovering	N/A	37	Culbertson et al. 2007
FLORIDA	1969	No. 2 Fuel Oil	557	Barge	Marine	MCR	Recovering	N/A	38	Culbertson et al. 2008a
FLORIDA	1969	No. 2 Fuel Oil	557	Barge	Marine	SED	Recovering	N/A	5	Michael et al. 1975, Sanders et al. 1980
FLORIDA	1969	No. 2 Fuel Oil	557	Barge	Marine	SED	Uncertain	N/A	20	Teal et al. 1992
FLORIDA	1969	No. 2 Fuel Oil	557	Barge	Marine	SED	None	N/A	30	Reddy et al. 2002
FLORIDA	1969	No. 2 Fuel Oil	557	Barge	Marine	VEG	Recovering	N/A	37	Culbertson et al. 2008b
John Heinz Wildlife Refuge	2000	Crude	650	Pipeline	Freshwater	REP	Recovered	0.4166	0.833	Saba and Spotila 2003
John Heinz Wildlife Refuge	2000	Crude	650	Pipeline	Freshwater	REP	None	N/A	3	Bell et al. 2006

Oil Spill Name	Year	Product Type (general)	Volume (metric tons) total	Platform	Environments	VEC *	Recovery Status	Years to recover	Study Duration	References
Kalamazoo River	2010	Heavy crude	3,247	Pipeline	Freshwater	BRD	Insufficient Data	N/A	UNK	Enbridge Energy (In Prep)
Kalamazoo River	2010	Heavy crude	3,247	Pipeline	Freshwater	FSH	Recovering	N/A	UNK	Michigan DNR
Kalamazoo River	2010	Heavy crude	3,247	Pipeline	Freshwater	MCR	Insufficient Data	N/A	0.25	Badra 2011
Oil Spill Name	Year	Product Type (general)	Volume (metric tons) total	Platform	Environments	VEC *	Recovery Status	Years to recover	Study Duration	References
Kalamazoo River	2010	Heavy crude	3,247	Pipeline	Freshwater	MCR	Recovering	N/A	1	MDEQ 2012
Kalamazoo River	2010	Heavy crude	3,247	Pipeline	Freshwater	REP	Recovering	N/A	1	Enbridge Energy (In Prep)
Komi	1995	Crude	136,370	Pipeline	Freshwater	SHR	Insufficient Data	N/A	2	Owens et al. 1999
Mackenzie Delta	1972	Crude	266	Experimental	Terrestrial	VEG	Recovering	N/A	1	Bliss and Wein 1972, Wein and Bliss 1973, Ross et al. 1977
METULA	1974	Light crude	52,000	Tanker/Ship	Marine	SHR	Recovering	N/A	12	Owens et al. 1999
METULA	1974	Light crude	52,000	Tanker/Ship	Marine	SHR	Recovering	N/A	30.5	Owens et al. 2008
METULA	1974	Light crude	52,000	Tanker/Ship	Marine	VEG	Recovering	N/A	23	Owens et al. 1999
Milford Haven	1969	No. 6 Fuel Oil	ND	Refinery	Marine	SED	Recovering	N/A	22	Baker et al. 1993
Milford Haven	1969	No. 6 Fuel Oil	ND	Refinery	Marine	VEG	Recovered	15	22	Baker et al. 1993
Moose Jaw	1974	Crude	2,143	Pipeline	Terrestrial	VEG	Recovering	N/A	4	De Jong 1980
Mt. Baker	1972	No. 2 Fuel Oil	22	Storage Tank	Terrestrial	VEG	Recovering	N/A	7	Belsky 1982
Mt. Baker	1972	No. 2 Fuel Oil	22	Storage Tank	Terrestrial	VEG	Recovering	N/A	UNK	Belsky, 1975

Appendix (Continued)

Oil Spill Name	Year	Product Type (general)	Volume (metric tons) total	Platform	Environments	VEC *	Recovery Status	Years to recover	Study Duration	References
NELLA DAN	1987	No. 2 Fuel Oil & Lubricating Oil	125.00	Tanker/Ship	Marine	ALG	Recovering	N/A	7	Smith and Simpson 1998
NELLA DAN	1987	No. 2 Fuel Oil & Lubricating Oil	125.00	Tanker/Ship	Marine	MCR	Recovering	N/A	7	Smith and Simpson 1998
NESTUCCA	1988	No. 6 Fuel Oil	734	Barge	Marine	FSH	Recovered	2	UNK	Hay and McCarter 2012
NESTUCCA	1988	No. 6 Fuel Oil	734	Barge	Marine	FSH	Uncertain	N/A	0.4	Hay et al. 1995
Nipisi	1970 - 1972	Crude	8,182	Pipeline	Terrestrial	SOI	Recovering	N/A	25	Wang et al. 1998
Norman Wells	1972	Crude	7	Experimental	Terrestrial	VEG	Recovering	N/A	4	Hutchinson and Freeedman 1978, Belsky 1982
Norman Wells	1972	Crude	7	Experimental	Terrestrial	VEG	Recovering	N/A	5	Hutchinson and Freeedman 1978
Pine River	2000	Light crude	704	Pipeline	Freshwater	FSH	Recovering	N/A	12	Goldberg 2011
Pine River	2000	Light crude	704	Pipeline	Freshwater	MCR	Recovered	1	1	DePennart 2004
Plains Rainbow	1970-1972	Crude	ND	Pipeline	Terrestrial	SOI	Recovering	N/A	25	Wang et al. 1998
Prudhoe Bay 1972	1972	Crude	1.185	Pipeline	Terrestrial	VEG	None	N/A	6	McKendrick and Mitchell 1978
Prudhoe Bay 1972	1972	Crude	1.185	Pipeline	Terrestrial	VEG	Recovering	N/A	6	McKendrick and Mitchell 1978
Prudhoe Bay 1976	1976	Crude & No. 2 Fuel Oil	0.01	Experimental	Terrestrial	VEG	Recovering	N/A	UNK	Walker et al. 1978
Reedy River	1996	No. 2 Fuel Oil	3,057	Pipeline	Freshwater	FSH	Recovered	4.3	9.3	Kubach et al. 2011
SEA EMPRESS	1996	Crude & No. 6 Fuel Oil	72,480	Tanker/Ship	Marine	ALG	Recovered	1	UNK	Moore 2006
SEA EMPRESS	1996	Crude & No. 6 Fuel Oil	72,480	Tanker/Ship	Marine	MCR	Recovered	6	UNK	Moore 2006; Nikitik and Robinson 2003, Rutt et al. 1998

Oil Spill Name	Year	Product Type (general)	Volume (metric tons) total	Platform	Environments	VEC *	Recovery Status	Years to recover	Study Duration	References
SEA EMPRESS	1996	Crude & No. 6 Fuel Oil	72,480	Tanker/Ship	Marine	SHR	Recovered	5	UNK	Moore 2006
Selendang Ayu	2004	No. 6 Fuel Oil Diesel	1,072	Tanker/Ship	Marine	BRD	Insufficient Data	N/A	0.33	Brewer et al. 2005
St. Lawrence River	1999	Crude	ND	Experimental	Freshwater	FSH	Recovering	N/A	1.25	Hodson et al. 2002
St. Lawrence River	1999	Crude	ND	Experimental	Freshwater	SED	Recovering	N/A	1.25	Hodson et al. 2002
Tallgrass Prairie	1993 - 2002	Crude	84,549	Experimental	Terrestrial	SOI	Recovering	N/A	2.3	Sublette et al. 2007
Tennessee Ponds	1981	Crude	ND	Experimental	Freshwater	MCR	Recovering	N/A	0.91	Cushman and Goyart 1984
TORREY CANYON	1967	Light crude	118,000	Tanker/Ship	Marine	VEG	Recovering	N/A	4	Burk 1977
TSESIS	1977	No. 5 Fuel Oil & No. 6 Fuel Oil	1,000	Tanker/Ship	Marine	MCR	Recovered	0.01	0.1	Johansson et al. 1980
TSESIS	1977	No. 5 Fuel Oil & No. 6 Fuel Oil	1,000	Tanker/Ship	Marine	MCR	Recovering	N/A	3	Fukuyama et al. 2000
TSESIS	1977	No. 5 Fuel Oil & No. 6 Fuel Oil	1,000	Tanker/Ship	Marine	MCR	Recovering	N/A	UNK	Linden et al. 1979
Tulita	1988	Crude Oil	3	Pipeline	Terrestrial	VEG	Recovering	N/A	3	Seburn et al. 1996
Wabamun Lake	2005	No. 6 Fuel Oil	126	Rail	Freshwater	FSH	None	N/A	0.75	Debruyn et al. 2007
Wabamun Lake	2005	No. 6 Fuel Oil	126	Rail	Freshwater	FSH	Insufficient Data	N/A	0.91	Transport Safety Board of Canada 2007
Wabamun Lake	2005	No. 6 Fuel Oil	126	Rail	Freshwater	SED	None	N/A	0.15	Zrum and Sergy 2006
Wabamun Lake	2005	No. 6 Fuel Oil	126	Rail	Freshwater	VEG	Recovering	N/A	2	Wernick et al. 2009
Wabamun Lake	2005	No. 6 Fuel Oil	126	Rail	Freshwater	WQL	Uncertain	N/A	0.11	Anderson 2006

* ALG = Algae, BRD = Birds, FSH = Fish, GWR = Groundwater, MCR = Macroinvertebrates, MML = Mammals, MCB = Microbial Community, REP = Reptiles, SED = Sediment, SHR = Shoreline, SOI = Soil, VEG = Vegetation, WQL = Water quality

Appendix References

Anderson, A.-M. (2006). Wabamun Lake Oil Spill August 2005: Data Report for Water and Sediment Quality in the Pelagic Area of the Lake (August 4-5 to September 15, 2005) (E. A. Division, Trans.): *Alberta Environment.*

Baca, B. J., Lankford, T. E., & Gundlach, E. R. (1987). Recovery of Brittany Coastal Marshes in the Eight Years Following the Amoco Cadiz Incident. *Paper presented at the 1987 International Oil Spill Conference.*

Badra, P. (2011). *Mussel Shell Survey Report: Kalamazoo River Unionid Mussel Shell survey in the Marshall and Battle Creek Area October 2010.* Lansing, MI.

Baker, J. M., Guzman, L. M., Bartlett, P. D., Little, D. I., & Wilson, C. M. (1993). Long-term fate and effects of untreated thick oil deposits on salt marshes. Paper presented at the International Oil Spill Conference, Washington, D.C.

Bell, B., Spotila, J. R., & Congdon, J. (2006). High incidence of deformity in aquatic turtles in the John Heinz National Wildlife Refuge. [Research Support, Non-U.S. Gov't]. *Environmental Pollution*, 142(3), 457-465. doi: 10.1016/j.envpol.2005.10.020.

Belsky, J. (1975). An oil spill in an alpine habitat. *Northwest Science*, 49, 141-146.

Belsky, J. (1982). Diesel oil spill in a subalpine meadow: 9 years of recovery. *Canadian Journal of Botany*, 60, 906-910.

Bliss, L. C., & Wein, R. W. (1972). Plant community responses to disturbances in the Western Canadian Arctic. *Journal of Canadian Botany,* 50, 1097-1109.

Bodkin, J. L., Ballachey, B. E., Dean, T. A., Fukuyama, A. K., Jewett, S. C., McDonald, L., VanBlaricom, B. R. (2002). Sea otter population status and the process of recovery from the 1989 'Exxon Valdez' oil spill. *Marine Ecology Progress Series* 241, 237-253.

Boehm, P. D., Neff, J. M., & Page, D. S. (2007). Assessment of polycyclic aromatic hydrocarbon exposure in the waters of Prince William Sound after the Exxon Valdez oil spill: 1989-2005. *Marine Pollution Bulletin*, 54, 339-356. doi: 10.1016/ j.marpolbul.2006.11.025.

Bowman, T. (1999). Bald eagle (Haliaeetus leucocephalus). In U. F. a. W. Service (Ed.), Exxon Valdez Oil Spill Trustee Council Restoration *Notebook*. Anchorage, AK.

Braddock, J. F., Lindstrom, J. E., & Prince, R. C. (2003). Weathering of a subarctic oil spill over 25 years: the Caribou-Poker Creeks Research Watershed experiment. *Cold Regions Science and Technology*, 36, 11-23. doi: 10.1016/s0165-232x(02)00076-9.

Brewer, R. (2005). The Selendang Ayu Oil Spill: Lessons Learned, conference proceedings, August 16-19, 2005, Unalaska, Alaska. *Paper presented at the Conference Proceedings August* 16-19, 2005- Unalaska, Alaska, Alaska.

Brown, E. D., Norcross, B. L., & Short, J. W. (1996). Introduction to Studies on the Effects of the (Exxon Valdez) Oil Spill on Early Life History Stages of Pacific Herring, (Clupea pallasi), in Prince William Sound, Alaska. *Canadian Journal of Fisheries and Aquatic Sciences* 53, 2337-2342.

Burk, J. C. (1977). A four year analysis of vegetation following an oil spill in a freshwater marsh. *Journal of Applied Ecology*, 14, 515-522.

California Department of Fish and Game. (2009). *Pacific Herring Commercial Fishing Regulations Final Supplemental Environmental Document.* Sacramento, CA.

California Department of Fish and Game. (2010). *Pacific Herring Commercial Fishing Regulations Final Supplemental Environmental Document.* Sacramento, CA.

California Department of Fish and Game. (2011). *Pacific Herring Commercial Fishing Regulations Final Supplemental Environmental Document.* Sacramento, CA.

Chabreck, R. H. (1973). Bird use of marsh ponds subjected to oil spills. *Paper presented at the Proceedings of the Louisiana Academy of Sciences.*

Collier, T. K., Krahn, M. M., Krone, C. A., Johnson, L. L., Myers, M. S., Chan, S.-L., & Varanasi, U. (1993). Oil exposure and the effects in subtidal fish following the Exxon Valdez oil spill. *Paper presented at the 1993 International Oil Spill Conference.*

Collins, C. M., Racine, C. H., & Walsh, M. E. (1994). The physical, chemical, and biological effects of crude oil spills after 15 years on a black spruce forest, Interior Alaska. *ARCTIC*, 47(2), 164-175.

Crunkilton, R. L., & Duchrow, R. M. (1990). Impact of a massive crude oil spill on the invertebrate fauna of a Missouri Ozark stream. *Environmental Pollution,* 63, 13-31. doi: 10.1016/0269-7491(90)90100-q.

Culbertson, J. B., Valiela, I., Peacock, E. E., Reedy, C. M., Carter, A., & VanderKruik, R. (2007). Long-term biological effects of petroleum residues on fiddler crabs in salt marshes. *ScienceDirect,* 54, 955-962.

Culbertson, J. B., Valiela, I., Olsen, Y. S., & Reedy, C. M. (2008a). Effect of field exposure to 38-year-old residual petroleum hydrocarbons on growth, condition index, and filtration rate of the ribbed mussel, Geukensia demissa. *ScienceDirect,* 154, 312-319.

Culbertson, J. B., Valiela, I., Pickart, M., Peacock, E. E., & Reedy, C. M. (2008b). Long-term consequences of residual petroleum on salt marsh grass. *Journal of Applied Ecology*, 45, 1284-1292. doi: 10.1111/j.1365-2664.2008.01477.x.

Cushman, R. M., & Goyert, J. C. (1984). Effects of a Synthetic Crude Oil on Pond Benthic Insects. *Environmental Pollution Series A*, 33, 163-186.

Dauvin, J.-C. (1998). The fine sand Abra alba community of the Bay of Morlaix twenty years after the Amoco Cadiz oil spill. *Marine Pollution Bulletin*, 36(9), 669-676.

Dauvin, J.-C. (2000). The Muddy Fine Sand Abra alba–Melinna palmata Community of the Bay of Morlaix Twenty Years After the Amoco Cadiz Oil Spill. *Marine Pollution Bulletin,* 40(6), 528-536. doi: 10.1016/s0025-326x(99)00242-8

De Jong, E. (1980). The effect of a crude oil spill on cereals. *Environmental Pollution Series A*, 22, 187-196.

De Pennart, H., Crowther, R., Taylor, T., Morden, M., & Mattison, S. (2004). The Use of Ecological Risk Assessment For Regional Management Of Aquatic Impacts. *Paper presented at the Environmental Services Association of Alberta Symposium,* Edmonton, AB.

Dean, T. A., Stekoll, M. S., Jewett, S. C., Smitha, R. O., & Hose, J. E. (1998). Eelgrass (Zostera marina L) in Prince William Sound, Alaska: effects of the Exxon Valdez oil spill. *Marine Pollution Bulletin*, 36(3), 201-210.

Debruyn, A. M. H., Wernick, B. G., Stefura, C., McDonald, B. G., Rudolph, B.-L., Patterson, L., & Chapman, P. M. (2007). In situ experimental assessment of lake whitefish development following a freshwater oil spill. *Environmental Science & Technology,* 41(20), 6983-6989.

Driskell, W. B., Fukuyama, A. K., Houghton, J. P., Lees, D. C., Shigenaka, G., & Mearns, A. J. (1993). Impacts on Intertidal Infauna: Exxon Valdez Oil Spill and Cleanup. *Paper presented at the 1993 International Oil Spill Conference.*

Enbridge Energy. (In Preparation-a). *Wildlife Response Report: Birds Enbridge Line 6b Spill Response.* Kalamazoo River, Michigan Marshall, Michigan.

Enbridge Energy. (In Preparation-b). *Wildlife Response Report: Turtles.* Enbridge Line 6b Spill Response. Kalamazoo River, Michigan. Marshall, Michigan,.

Esler, D., & Iverson, S. A. (2010). Female Harlequin Duck Winter Survival 11 to 14 Years After the Exxon Valdez Oil Spill. *Journal of Wildlife Management,* 74(3), 471-478. doi: 10.2193/2008-552.

Esler, D., Trust, K. A., Ballachey, B. E., Iverson, S. A., Lewis, T. L., Rizzolo, D. J., . . . Wilson, B. W. (2010). Cytochrome P4501A Biomarker Indication of Oil Exposure in Harlequin Ducks up to 20 Years after the Exxon Valdez Oil Spill. [Article]. *Environmental Toxicology and Chemistry,* 29(5), 1138-1145. doi: 10.1002/etc.129.

Essaid, H. I., Bekins, B. A., Herkelrath, W. N., & Delin, G. N. (2011). Crude Oil at the Bemidji Site: 25 Years of Monitoring, Modeling, and Understanding. [Article]. *Ground Water*, 49(5), 706-726. doi: 10.1111/j.1745-6584.2009.00654.x.

Exxon Valdez Oil Spill Trustee Council. (2002). Exxon *Valdez Oil Spill Restoration Plan Update on Injured Resources and Services Exxon Valdez Oil Spill Trustee Council.* Anchorage, AK.

Exxon Valdez Oil Spill Trustee Council. (2010). *Exxon Valdez Oil Spill Restoration Plan 2010 Update Injured Resources and Services Exxon Valdez Oil Spill Trustee Council.* Anchorage, AK.

Fitzgerald, D. (2007-2008). Nutrient Cycling, Microbes and the Fate of Oil. *Agroborealis,* 39(2), 20-25.

Fukuyama, A. K., Shigenaka, G., & Hoff, R. Z. (2000). Effects of Residual Exxon Valdez Oil on Intertidal Protothaca staminea: Mortality, Growth, and Bioaccumulation of Hydrocarbons in Transplanted Clams. *Marine Pollution Bulletin*, 40(11), 1042-1050.

Gilfillan, E. S., & Vandermeulen, J. H. (1978). Alterations in Growth and Physiology of Soft-Shell Clams, Mya arenaria, Chronically Oiled with Bunker C from Chedabucto Bay, Nova Scotia, 1970-76. *Journal of the Fisheries Research Board of Canada,* 35, 630-636.

Glemarec, M., & Hussenot, E. (1982). A three-year ecological survey in Benoit and Wrac'h Abers following the Amoco Cadiz oil spill. *Netherlands Journal of Sea Research*, 16, 483-490.

Goldberg, H. (2011). *Pine River 2011 Fisheries Update: Status of Recovery Post-2000 Pipeline Rupture.* Calgary, AB.

Hampton, S., Montalvo, A., Higgins, D., & Sollberger, P. (2002). *Assessment of Natural Resource Damages as a Result of the East Walker River Oil Spill on December 30*, 2000. Sacramento, CA.

Hay, D. E., & McCarter, P. B. (2012). *Herring Spawning Areas of British Columbia: A Review, Geographical Analysis and Classification.* Nanaimo, British Columbia: Retrieved from http://www.pac.dfo-mpo.gc.ca/science/ species-especes/pelagic-pelagique/herring-hareng/herspawn/pages/ acknowl1-eng.htm.

Hay, D. E., McCarter, P. B., & D.Grant. (1995). Nestucca Oil Spill and Impacts on Herring Eggs. *Paper presented at the Proceedings of the Seventh Pacific Coast Herring*

Workshop. Available online: http://www.pac.dfo-mpo.gc.ca/science/species-especes/pelagic-pelagique/herring-hareng/herspawn/pages/acknowl1-eng.htm.

Hayes, M. O., & Michel, J. (1999). Factors Determining the Long Term Persistence of Exxon Valdez Oil in Gravel Beaches. *Marine Pollution Bulletin*, 38(2), 92-101.

Higgins, D. K. (2002). *Assessment of Damages to Natural Resources in the East Walker River from the Advanced Fuel Filtration Spill: Water, Sediment, and Fish Tissue Analysis* (N. F. a. W. Office & D. o. E. Quality, Trans.). Reno, NV: U.S. Fish and Wildlife Service.

Hodson, P. V., Ibrahim, I., Zambon, S., Ewert, A., & Lee, K. (2002). Bioavailability to Fish of Sediment PAH as an Indicator of the Success of In Situ Remediation Treatments at an Experimental Oil Spill. [Article]. *Bioremediation Journal*, 6(3), 297.

Hoff, R. Z., & Shigenaka, G. (1999). Lessons from Ten Years of Post-Exxon Valdez Monitoring on Intertidal Shorelines. *Paper presented at the International Oil Spill Conference*, Seattle, WA.

Hoff, R. Z., Shigenaka, G., & Henry, C. B., Jr. (1993). *Salt Marsh Recovery from a Crude Oil Spill: Vegetation, Oil Weathering, and Response. Paper presented at the 1993 International Oil Spill Conference*, Tampa Bay, FL.

Houghton, J. P., Gilmour, R. H., Lees, D. C., Driskell, W. B., Lindstrom, S. C., & Mearns, A. (1997). Prince William Sound Intertidal Biota Seven Years Later: Has it Recovered? . *Paper presented at the 1997 International Oil Spill Conference*, Ft. Lauderdale, FL.

Humphrey, B., Owens, E. H., & Sergy, G. (1991). Long-term Results from the BIOS Shoreline Experiment- Surface Oil Cover. *Paper presented at the 1991 International Oil Spill Conference*, San Diego, CA.

Hutchinson, T. C., & Freedman, W. (1978). Effects of Experimental Crude Oil Spills on Subarctic Boreal Forest Vegetation Near Norman Wells, N.W.T., Canada. *Canadian Journal of Botany*, 56, 2424-2433. doi: 10.1139/b78-294

Irvine, G. V., Mann, D. H., & Shorts, J. W. (1999). Multi-year Persistence of Oil Mousse on High Energy Beaches Distant from Exxon Valdez Spill Origin. *Marine Pollution Bulletin*, 38(7), 572-584.

Jenkins, T. F., Johnson, L. A., Collins, C. M., & McFadden, T. T. (1978). The Physical, Chemical and Biological Effects of Crude Spills on Black Spruce Forest, Interior Alaska. *Arctic*, 31, 305-323.

Jewett, S. C., Dean, T. A., Woodin, B. R., Hoberg, M. K., & Stegeman, J. J. (2002). Exposure to hydrocarbons 10 years after the Exxon Valdez Oil spill: Evidence from Cytochrome P4501A Expression and Biliary FACs in Nearshore Demersal Fishes. [Article]. *Marine Environmental Research,* 54, 21-48.

Johansson, S., Larsson, U., & Boehm, P. (1980). The Tsesis Oil Spill Impact on the Pelagic Ecosystem. [Article]. *Marine Pollution Bulletin*, 11(10), 284-293.

Kingston, P. F. (1999). Recovery of the Marine Environment Following the Braer spill, Shetland. *Paper presented at the 1999 International Oil Spill Conference*, Seattle, Washington

Kingston, P. F. (2002). Long-term Environmental Impact of Oil Spills. [Review Paper]. Spill Science & Technology Bulletin, 7(1-2), 53-61.

Krebs, C. T., & Burns, K. A. (1977). Long-Term Effects of an Oil Spill on Populations of the Salt-Marsh Crab Uca pugnax. *Science,* 197(4302), 484-487.

Kubach, K. M., Scott, M. C., & Bulak, J. S. (2011). Recovery of a Temperate Riverine Fish Assemblage from a Major Diesel Oil spill. [Article]. *Freshwater Biology*, 56, 503-518. doi:10.1111/j.1365-2427.2010.02517.x.

Lance, B. K., McDonald, L. L., Irons, D. B., & Kendall, S. J. (2001). An Evaluation of Marine Bird Population Trends Following the Exxon Valdez Oil Spill, Prince William Sound, Alaska. [Article]. *Marine Pollution Bulletin,* 42(4), 298.

Lee, K., Prince, R. C., Greer, C. W., Doe, K. G., Wilson, J. E. H., Cobanli, S. E., . . . Tremblay, G. H. (2003). Composition and Toxicity of Residual Bunker C Fuel Oil in Intertidal Sediments After 30 Years. *Spill Science & Technology Bulletin*, 8(2), 187-199. doi: 10.1016/s1353-2561(03)00014-8.

Lee, K., Wohlgeschaffen, G. D., Tremblay, G. H., Vandermeulen, J. H., Mossman, D. C., Doe, K. G., . . . Haith, C. E. (1999). *Natural Recovery Reduces Impact of the 1970 Arrow Oil Spill. Paper presented at the 1999 International Oil Spill Conference,* Seattle, WA.

Lee, R. F., & Page, D. S. (1997). Petroleum Hydrocarbons and their Effects in Subtidal Regions after Major Oil Spills. *Marine Pollution Bulletin*, 34(11), 928-940. doi: 10.1016/s0025-326x(97)00078-7

Linden, O., Elmgren, R., & Boehm, P. (1979). The Tsesis Oil Spill: its Impact on the Coastal Ecosystem of the Baltic Sea. *Ambio*, 8(6), 246-253.

Lindstrom, J. E., Barry, R. P., & Braddock, J. F. (1999). Long-term Effects on Microbial Communities after a Subarctic Oil Spill. *Soil Biology and Biochemistry*, 31, 1677-1689. doi: 10.1016/s0038-0717(99)00081-4

Llanso, R. J., & Volstad, J. (2001). *Patuxent River Oil Spill: Assessment of Impacts on Benthos Final Report.* 9200 Rumsey Road Columbia, Maryland 21045: Swanson Creek Natural Resource Damage Assessment Trustee Council.

Lytle, D. A., & Peckarsky, B. L. (2001). Spatial and Temporal Impacts of a Diesel Fuel Spill on Stream Invertebrates. [Article]. *Freshwater Biology,* 46, 693-704. doi: 10.1046/j.1365-2427.2001.00695.x

MacCarone, A. D., & Brzorad, J. N. (2000). Wading Bird Foraging: Response and Recovery from an Oil Spill. Waterbirds: *The International Journal of Waterbird Biology*, 23(2), 246-257.

McKendrick, J. D., & Mitchell, W. W. (1978). Fertilizing and seeding oil-damaged Arctic tundra to effect vegetation recovery Prudhoe Bay, Alaska. *ARCTIC,* 31(3), 296-304.

Michael, A. D., Van Raalte, C. R., & Brown, L. S. (1975). Long-term Effects of an Oil Spill at West Falmouth, Massachusetts. *Paper presented at the Conference on Prevention and Control of Oil Pollution*, Washington, D.C.

Michel, J., & Hayes, M. O. (1993). Persistence and Weathering of Exxon Valdez Oil in the Intertidal Zone--3.5 Years Later. *Paper presented at the 1993 International Oil Spill Conference*, Washington, D.C.

Michel, J., Nixon, Z., Dahlin, J., Betenbaugh, D., White, M., Burton, D., & Turley, S. (2009). Recovery of Interior Brackish Marshes Seven Years after the Chalk Point Oil Spill. [Research Support, U.S. Gov't, Non-P.H.S.]. *Marine Pollution Bulletin*, 58, 995-1006. doi: 10.1016/j.marpolbul.2009.02.015.

Michel, J., Nixon, Z., Hayes, M. O., Shorth, J., Irvine, G., Betenbaugh, D., . . . Mann, D. (2010). *Distribution of Subsurface Oil from the Exxon Valdez Oil Spill* (National Oceanic and Atmospheric Administration, Trans.). Juneau, AK.

Michigan Department of Environmental Quality. (2012). *A biological survey of sites on the Kalamazoo River and Talmadge Creek near the Enbridge oil spill in Marshall Calhoun County*, Michigan. Michigan Department of Environmental Quality (MDEQ).

Michigan Department of Natural Resources. (2011). *A Fish Survey of Sites on The Kalamazoo River and Talmadge Creek near the Enbridge Oil Spill in Marshall Calhoun and Kalamazoo Counties*, Michigan September 2010. Lansing, MI.

Moore, J. (2006). Long term ecological impacts of marine oil spills. *Paper presented at the Interspill 2006 Conference*, London ExCeL.

Moore, J., Taylor, P., & Hiscock, K. (1995). Rocky Shores Monitoring Programme. *Paper presented at the Proceedings of the Royal Society of Edinburgh*, Edinburgh, Scotland.

Moore, J. J. (2006). State of the marine environment in South West Wales 10 *Years after the Sea Empress Oil Spill* (L. a. M. Coastal Assessment, Cosheston, Pembrokeshire, Trans.). In W. G. Sanderson (Ed.), Countryside Council for Whales (pp. 24 plus vi). Cosheston, Pembrokeshire, Whales, UK: Coastal Assessment, Liaison and Monitoring.

Murphy, S. M., & Mabee, T. J. (2000). Status of Black Oystercatchers in Prince William Sound, Alaska, Nine Years After the Exxon Valdez Oil Spill. *Waterbirds: The International Journal of Waterbird Biclogy*, 23(2), 204-213.

Neff, J. M., & Burns, W. A. (1996). Estimation of Polycyclic Aromatic Hydrocarbon Concentrations in the Water Column Based on Tissue Residues in Mussels and Salmon: an Equilibrium Partitioning Approach. [Article]. *Environmental Toxicology and Chemistry*, 15(12), 2240-2251.

Neff, J. M., Owens, E. H., Stoker, S. W., & McCormick, D. M. (1995). Shoreline Oiling Conditions in Prince William Sound Following the Exxon Valdez Oil Spill. *Philadelphia: American Society for Testing and Materials* (ASTM STP 1219).

Nikitik, C. C. S., & Robinson, A. W. (2003). Patterns in Benthic Populations in the Milford Haven Waterway following the 'Sea Empress' Oil Spill with Special Reference to Amphipods. [Article]. *Marine Pollution Bulletin*, 46, 1125-1141. doi: 10.1016/s0025-326x(03)00236-4

Owens, E. H. (1978). Mechanical Dispersal of Oil Stranded in the Littoral Zone. *Journal of the Fisheries Research Board of Canada*, 35, 563-572.

Owens, E. H., Humphrey, B., Hope, D., Robson, W., & Harper, J. R. (1987). The Fate of Stranded Oil Four Years after an Experimental Spill on a Sheltered Gravel Beach. *Paper presented at the 1987 International Oil Spill Conference,* Washington, D.C.

Owens, E. H., Sienkiewiez, M. A., & Sergy, G. A. (1999). Evaluation of Shoreline Ceaning Vrsus Natural Recovery: the Metula Spill and Komi Operations. *Paper presented at the 1999 International Oil Spill Conference,* Seattle, WA.

Owens, E. H., Taylor, E., & Humphrey, B. (2008). The Persistence and Character of Stranded Oil on Coarse-Sediment Beaches. [Research Support, Non-U.S. Gov't Review]. *Marine Pollution Bulletin*, 56, 14-26. doi: 10.1016/j.marpolbul.2007.08.020.

Page, D. S., Boehm, P. D., Brown, J. S., Neff, J. M., Burns, W. A., & Bence, A. E. (2005). Mussels document loss of bioavailable polycyclic aromatic hydrocarbons and the return to baseline conditions for oiled shcrelines in Prince William Sound, Alaska. [Research Support, Non-U.S. Gov't]. *Marine Environmental Research*, 60(4), 422-436. doi: 10.1016/j.marenvres.2005.01.002.

Parsons, K. C. (1996). Recovering From Oil Spills: The Role of Proactive Science in Mitigating Adverse Effects. *Colonial Waterbirds*, 19(1), 149-153.

Pearson, W. H., Moksness, E., & Skalski, J. R. (1995). *A Field and Laboratory Assessment of Oil Spill Effects on Survival and Reproduction of Pacific Herring Following the Exxon Valdez Spill.* Philadelphia: American Society for Testing and Materials.

Poulton, B. C., Callahan, E. V., Hurtubise, R. D., & Mueller, B. G. (1998). Effects of an Oil Spill on Leafpack- Inhabiting Macroinvertebrates in the Chariton River Missouri. *Environmental Pollution*, 99, 115-122.

Prince, R. C., Garrett, R. M., Bare, R. E., Grossman, M. J., Townsend, T., Suflita, J. M., . . . Lessard, R. R. (2003). The Roles of Photooxidation and Biodegradation in Long-term Weathering of Crude and Heavy Fuel Oils. *Spill Science & Technology Bulletin*, 8(2), 145-156. doi: 10.1016/s1353-2561(03)00017-3.

Prince, R. C., Owens, E. H., & Sergy, G. A. (2002). Weathering of an Arctic Oil Spill over 20 Years: the BIOS experiment revisited. [Article]. *Marine Pollution Bulletin,* 44, 1236-1242.

Racine, C. H. (1994). Long-term Recovery of Vegetation on Two experimental Crude Oil Spills in Interior Alaska Black Spruce Taiga. *Canadian Journal of Botany*, 72, 1171-1177. doi: 10.1139/b94-143.

Reddy, C. M., Eglinton, T. I., Hounshell, A., White, H. K., Li, X., Gaines, R. B., & Frysinger, G. S. (2002). The West Falmouth Oil Spill after Thirty Years: The Persistence of Petroleum Hydrocarbons in Marsh Sediments. [Article]. *Environmental Science & Technology*, 36(22), 4754-4760.

Revesz, K., Coplen, T. B., Baedecker, M. J., Glynn, P. D., & Hult, M. (1995). Methane Production and Consumption Monitored by Stable H and C Isotope Ratios at a Crude Oil Spill Site, Bemidji, Minnesota. *Applied Geochemistry,* 10, 505-516. doi: 10.1016/0883-2927(95)00021-6.

Ross, S. L., Logan, W. J., & Rowland, W. (1977). *The Beaufort Sea and the Search for Oil: Oil Spill Countermeasures.* Ottawa, CA: Beaufort Sea Project.

Rutt, G. P., Levell, D., Hobbs, G., Rostron, D. M., Bullimore, B., Law, R. J., & Robinson, A. W. (1998, February 11-13 1998). The Effect on the Marine Benthos. *Paper presented at the Sea Empress Oil Spill: Proceedings of the International Conference*, Cardiff.

Saba, V. S., & Spotila, J. R. (2003). Survival and Behavior of Freshwater Turtles after Rehabilitation from an Oil Spill. [Article]. *Environmental Pollution,* 126, 213-223. doi: 10.1016/s0269-7491(03)00192-1.

Sanders, H. L., Grassle, F. J., Hampson, G. R., Morse, L. S., Garner-Price, S., & Jones, C. C. (1980). Anatomy of an Oil Spill: Long-Term Effects from the Grounding of the Barge Florida off West Falmouth, Massachusetts. *Journal of Marine Research*, 38(2), 265-380.

Seburn, D. C., Kershaw, G. P., & Kershaw, L. J. (1996). Vegetation Response to a Subsurface Crude Oil Spill on a Subarctic Right-of-Way, Tulita (Fort Norman), Northwest Territories, Canada. *Arctic,* 49(4), 321-327.

Seymour, R. J., & Geyer, R. A. (1992). Fate and Effects of Oil Spills. *Annual Reviews of Energy Environment,* 17, 261-283.

Short, J. W., Lindeberg, M. R., Harris, P. M., Maselko, J. M., Pella, J. J., & Rice, S. D. (2003). Evaluation of Oil Remaining in Prince William Sound from the Exxon Valdez Oil Spill (A. F. S. C. National Marine Fisheries Service, NOAA, Trans.) *Exxon Valdez Oil Spill Restoration Project Final Report.* Auke Bay Laboratory, AK.

Smith, S. D. A., & Simpson, R. D. (1998). Recovery of benthic communities at Macquarie Island (sub-Antarctic) following a small oil spill. [Article]. *Marine Biology*, 131, 567-581. doi: 10.1007/s002270050349.

Sparrow, E. B., Davenport, C. V., & Gordon, R. C. (1978). Response of Microorganisms to Hot Crude Oil Spills on a Subarctic Taiga Soil. *Arctic* 31(3), 324-338.

Sparrow, S. D., & Sparrow, E. B. (1988). Microbial Biomass and Activity in a Subarctic Soil Ten Years After Crude Oil Spills. *Journal of Environmental Quality,* 17(2), 304-309.

Spooner, M. F. (1978). Editorial Introduction Amoco Cadiz oil spill. *Marine Pollution Bulletin,* 9(11), 281-284.

Stagg, R. M., Robinson, C., McIntosh, A. M., Moffat, C. F., & Bruno, D. W. (1998). The effects of the 'Braer' oil spill, Shetland Isles, Scotland, on P4501A in Farmed Atlantic Salmon (Salmo salar) and the Common Dab (Limanda limanda). [Article]. *Marine Environmental Research*, 46(1-5), 301-306.

Stoker, S. W., Neff, J., M., Schroeder, T. R., & McCormick, D. M. (1993). Biological Conditions of Shorelines following the Exxon Valdez Spill. *Paper presented at the 1993 International Oil Spill Conference*.

Sublette, K., Jennings, E., Mehta, C., Duncan, K., Brokaw, J., Todd, T., & Thoma, G. (2007). Monitoring Soil Ecosystem Recovery Following Bioremediation of a Terrestrial Crude Oil Spill With and Without a Fertilizer Amendment. [Article]. *Soil & Sediment Contamination*, 16, 181-208. doi: 10.1080/15320380601166470.

Teal, J. M., Farrington, J. W., Burns, K. A., Stegeman, J. J., Tripp, B. W., Woodin, B., & Phinney, C. (1992). The West Falmouth Oil Spill after 20 Years: Fate of Fuel Oil Compounds and Effects on Animals. *Marine Pollution Bulletin,* 24(12), 607-614. doi: 10.1016/0025-326x(92)90281-a.

Thomas, M. L. H. (1978). Comparison of Oiled and Unoiled Intertidal Communities in Chedabucto Bay, Nova Scotia. *Journal of the Fisheries Research Board of Canada*, 35, 707-716.

Transportation Safety Board of Canada. (2007). Railway Investigation Report Derailment Canadian National Freight Train M30351-03 Mile 49.4, Edson Subdivision Wabamun, Alberta 03 August 2005 *Report Number* R05E0059. Gatineau, QC.

Trust, K. A., Esler, D., Woodin, B. R., & Stegeman, J. J. (2000). Cytochrome P450 1A Induction in Sea Ducks Inhabiting Nearshore Areas of Prince William Sound, Alaska. *Marine Pollution Bulletin,* 40(5), 397-403. doi: 10.1016/s0025-326x(99)00236-2.

United States Geological Survey (USGS), & Minnesota Water Science Center. (2012). *Bemidji Crude-Oil Research Project Retrieved* July, 2012, from http://mn.water.usgs.gov /projects/bemidji/index.html.

University of Pittsburgh. (1990). Assessment of environmental effects from the January 2, 1988 diesel oil spill into the Monogahela River: *Final report on two year study effort* (C. f. H. M. Research, Trans.) (pp. 288). Pittsburgh, PA.

Venosa, A. D., & Zhu, X. (2003). Biodegradation of Crude Oil Contaminating Marine Shorelines and Freshwater Wetlands. *Spill Science & Technology Bulletin*, 8(2), 163-178. doi: 10.1016/s1353-2561(03)00019-7.

Walker, D. A., Webber, P. J., Everett, K. R., & Brown, J. (1978). Effects of Crude and Diesel Oil Spills on Plant Communities at Prudhoe Bay, Alaska, and the Derivation of Oil Spill Sensitivity Maps. *ARCTIC,* 31(3), 242-259.

Wang, Z., Fingas, M., Blenkinsopp, S., Sergy, G., Landriault, M., Sigouin, L., & Lambert, P. (1998). Study of the 25-yr old Nipisi Oil Spill: Persistence of Oil Residues and Comparisons between Surface and Subsurface Sediments. *Environmental Science Technology*, 32, 2222-2232.

Webster, L., McIntosh, A. M., Russell, M., Walsham, P., Packer, G., Dalgarno, E. J., & Moffat, C. F. (2005). *Composition and Concentration of Hydrocarbons in Sediment Samples Collected During 2003 from West Burra and St Magnus Bay*, Shetland Islands (M. Labratory, Trans.). Victoria Road, Aberdeen AB119DB: Fisheries Research Services.

Wein, R. W., & Bliss, L. C. (1973). Experiments Crude Oil Spills on Arctic Plant Communities. [Article]. *Journal of Applied Ecology,* 10(3), 671-682.

Wernick, B. G., DeBruyn, A. M., Patterson, L., & Chapman, P. M. (2009). Effects of an Oil Spill on the Regrowth of Emergent Vegetation in a Northern Alberta Lake. Archives of Environmental Contamination Toxicology, 57, 697-706. doi: 10.1007/s00244-009-9311-1

Wiens, J. A., Day, R. H., Murphy, S. M., & Parker, K. R. (2001). On Drawing Conclusions Nine Years After the Exxon Valdez Oil Spill. [Article]. *Condor*, 103(4), 886-892.

Zrum, L., & Sergy, G. (2006). Potential for Biodegradation of Sub-Littoral Residual Oil by Naturally Occurring Microorganisms following the Lake Wabamun Train Derailment (B. Sciences, Trans.). In J. Foght (Ed.): *Alberta Environment, Environment Canada*, University of Alberta.

In: Oil Spills
Editor: Adam Clifton

ISBN: 978-1-63321-515-3

Chapter 2

OBSERVATIONS AND ANALYSES OF MICROBIAL PATHOGENS AFFECTED BY THE DEEPWATER HORIZON OIL SPILL

Joong-Wook Park* and Kewei Yu
Department of Biological and Environmental Sciences
Troy University, Troy, AL, US

ABSTRACT

Much attention has been drawn to oil-degrading microorganisms in the Gulf of Mexico region in order to bioremediate the oil spills after the Deepwater Horizon accident in 2010, while little has been investigated in the impact of spilled oils on microbial pathogens in the Gulf region. This chapter summarizes the observations and analyses of microbial pathogens impacted by the Deepwater Horizon oil spill in the Gulf of Mexico region. In the observation section, microbial pathogens and their waterborne disease outbreaks in the United States are briefly reviewed. Especially in this section, two types of water are focused; tap water treatment procedure in relation to waterborne pathogens and recreational water as a route of disease transmission. Then, waterborne disease outbreaks after natural or anthropogenic disasters and abiotic factors affecting the fate of waterborne pathogens are discussed. At the end of this section, recent publications on the microbial pathogens in the Gulf region after the Deepwater Horizon oil spill are introduced, which are mainly focused on pathogenic marine *Vibrio* species that cause seafood poisonings and skin infections. In the analyses section, various molecular techniques are introduced to detect and monitor marine microbial pathogens including bacteria, protists, and viruses. Lastly, a brief chronicle of the U.S. government regulations to control waterborne disease outbreaks in the Gulf region are introduced, and several aspects to control future pathogen outbreaks in the Gulf region are discussed.

* Corresponding author: Email: jwpark@troy.edu.

1. INTRODUCTION

Marine and freshwater environments are often contaminated by natural and anthropogenic sources. Among all aspects of water contamination that require attention, waterborne diseases by microbial pathogens pose a major and direct threat to human health. In coastal regions, the risk of disease outbreaks tends to be increased significantly during or after natural disasters, such as tsunamis, hurricanes, and floods by running off sewage, garbage, and waste. Additionally these natural disasters can often accompany with oil spills since many oil-refining facilities are located along the coasts. Northern coastal region of the Gulf of Mexico has experienced numerous hurricanes and has the largest number of oil refineries in the United States (Figure 1), thus this region exposes a high risk of waterborne pathogen outbreak and oil spill at the same time.

Table 1. Etiology of waterborne outbreaks reported in the U.S., 1991-2002 (adapted from Craun et al., 2006)

Etiological agent	Outbreaks	Cases
Protists	**41**	**410,656**
Giardia	25	2,283
Cryptosporidium	15	408,371
Naegleria fowleri	1	2
Bacteria	**41**	**3,191**
Escherichia coli O157:H7	11	288
Shigella	9	663
Campylobacter jejuni	7	360
Legionella	6	80
Salmonella, non-Typhoid	3	833
Vibrio cholerae	2	114
Plesiomonas shigelloides	1	60
Etiological agent	**Outbreaks**	**Cases**
Campylobacter and *Yersinia*	1	12
Escherichia coli O157:H7 and *Campylobacter*	1	781
Viruses	**15**	**3,487**
Norovirus	12	3,361
Hepatitis A virus	2	56
Unidentified small round structured viruses (SRSV)	1	70
Acute gastroenteritis illness (AGI)	**77**	**16,036**
Chemical	**33**	**577**

The 2010 Deepwater Horizon (DH) oil spill brings a wide range of considerable changes affecting natural ecosystems. This catastrophic oil spill damaged properties and environments, devastated tourism and seafood industries, and caused health problems in the affected communities. Many studies have focused on the microbial bioremediation of the spilled oil with only a few studies conducted to evaluate the effects of the spilled oil on microbial pathogens. This chapter discusses general information of waterborne microbial pathogens in the U.S. and their fate in the oil-spilled Gulf of Mexico.

2. OBSERVATION

2.1. Microbial Pathogens and Waterborne Outbreaks in the United States

According to the U.S. Environmental Protection Agency (EPA), the United States has one of the safest water supplies in the world (EPA, 2011a). However, waterborne disease outbreaks have been prevalent in the United States (Table 1 and 2) (Craun et al., 2006; CDC MMWR, 2014). Most severe waterborne disease outbreaks in the U.S. have been caused by protists, unicellular eukaryotic microorganisms, including *Giardia*, *Cryptosporidium*, and *Naegleria*. Among those, *Cryptosporidium* is the most dominant microbial pathogens that frequently cause severe waterborne disease outbreaks (Table 1 and 2). Etiology of outbreaks caused by *Cryptosporidium* revealed 15 outbreaks with 408,371 cases during 1991 to 2001 with an average of 27,225 cases per outbreak (Table 1). In contrast, an average number of cases per bacterial outbreak was relatively low. For example, *Escherichia coli* O157:H7 outbreaks were 11 times with 288 cases with an average of 26 cases per outbreak during the same period (1991 to 2001) (Craun et al., 2006). Also, it is noticeable that bacterial outbreaks have been caused by a variety of species including *E. coli, Shigella, Campylobacter, Legionella, Salmonella, Vibrio, Plesiomonas, Yersinia,* and *Pseudomonas* (Table 1 and 2). Viral outbreaks were mainly caused by norovirus (Table 1 and 2), a highly contagious virus that can cause acute gastroenteritis.

People are exposed to tap water and recreational water that may carry microbial pathogens in their daily life. Tap water quality is strictly regulated and maintained in the U.S due to its direct human consumption. Five major processes of water treatment used by community water systems are (1) coagulation and flocculation, (2) sedimentation, (3) filtration, (4) disinfection, and (5) storage (Figure 2). At first, positively charged chemicals are added to neutralize negatively charged impurities. This process allows the particles to coagulate and form large clumps called "floc". The floc is precipitated and the top clear part of water passes through layers of filters. The filtered water is treated with chemical disinfectants (such as chlorine, chlorine dioxide, and ozone) or physical disinfectants (such as ultraviolet light, electronic radiation, and heat). The treated water is temporarily stored in a closed tank or reservoir before delivery (Figure 2). Through coagulation, flocculation, and sedimentation steps, 90 to 99 % of the bacteria, viruses, and protists can be removed (WHO, 2004). Filtration and disinfection can significantly remove and kill microbial pathogens by physical and chemical treatments. Storage step also contributes to deactivate microbial pathogens. Despite of all these efforts, microbial pathogens sometimes can pass through the water treatment process and cause waterborne disease outbreaks in the U.S. (Table 2). Since dormant forms of protists, such as oocysts of *Cryptosporidium* and cysts of *Giardia* and *Naegleria* are resistant to disinfection processes, they may remain in the treated tap water and could cause disease outbreaks (Table 2).

Waterborne disease outbreaks may also be originated from recreational water. Recreation water refers to swimming pools, hot tubs, Jacuzzis, spas, fountains, lakes, rivers, springs, ponds, streams, or oceans (CDC, 2010). Since a large portion of the recreational water is not processed by water treatment systems, there is a high risk of exposure to microbial pathogens. The microbial pathogens in the recreational water can directly contact and invade human body through any cuts, scrapes, or ingestion in the act of swimming, wading, surfing, diving, fishing, etc. A recent report by the Centers for Disease Control and Prevention (CDC) shows a tendency of steady increase of waterborne disease outbreaks associated with the recreational water (Figure 3) (CDC MMWR, 2014).

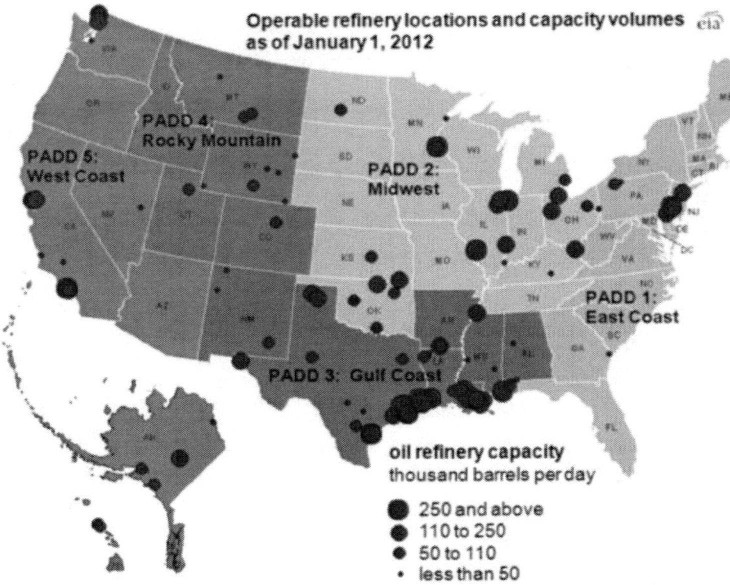

Figure 1. Operable refinery locations and capacity volumes as of January 1, 2012. (adapted from U.S. Energy Information Administration, Refinery Capacity Report, 2012).

2.2. Microbial Pathogens in the Gulf of Mexico

Microbial pathogens are widespread in various environments including land, ocean, freshwater, and even intestinal ducts of animals and humans (Ivnitski et al., 1999). In general, disease transmission and outbreaks caused by microbial pathogens tend to be amplified during or after natural and anthropogenic disasters (Smith, 2005). Tsunamis, hurricanes, and floods are examples of natural disasters that frequently cause significant environmental pollution in aquatic ecosystems. During and/or after the natural disasters, untreated and undertreated sewage, garbage, and waste are swept into the storm drains connected to the nearby aquatic ecosystems, and eventually cause water eutrophication and contamination. Especially, rapid growth of waterborne microbial pathogens can be a direct threat to public health. Furthermore, natural disasters can often accompany with oil spills since the major oil refineries are located along the coastline (U.S. EIA, 2012) such as the Gulf of Mexico (Figure 1).

Abiotic environmental factors such as temperature, moisture, nutrients, pH, and salinity have profound impacts on the fate of microorganisms (Oliver, 2009). Certain microbial pathogens are known to be able to survive in two distinctive environments - their hosts and natural environments. These pathogens respond to changes in surroundings by modulating gene expressions - activating necessary genes and inactivating unnecessary genes (Chowdhury et al., 1996). In aquatic environments, certain pathogenic bacteria are known to enter a physiological state called "viable but nonculturable (VBNC)" in response to environmental stresses for their survival (Caro et al., 1999; Oliver, 2009). *Legionella pneumophila, E. coli*, *Salmonella enteritidis*, *Shigella* spp., *Campylobacter* spp., and *Vibrio* spp. are among the pathogenic bacteria that can enter into the VBNC state (McKay, 1992, Mizunoe et al. 1999, Yaron & Matthews 2002). Temperatures (Besnard et al., 2002), nutrients (Cook & Bolster, 2007), osmotic pressures (Asakura et al., 2008), heavy metals (Ghezzi & Steck, 1999), and oxygen levels (Kana et al., 2008) are examples of environmental stresses that can induce the VBNC state (Oliver, 2009). Temperature also affects many other cellular processes (Hurme & Rhen, 1998; Muoghalu, 2003) and determines bacterial abundance, variability, production, and growth rate in the environment (Shiah & Ducklow, 1994). Several studies have demonstrated that temperature is the strongest environmental predictor in determining the dynamics of bacterial community (DePaola et al., 2003; Thompson et al., 2004: Zimmerman et al., 2007). For example, virulence gene cassettes of *Salmonellae*, *Shigellae*, *Yersinae*, *Bordetella pertusis*, *Borrelia burgdorferi*, and *Listeria monocytogens* are switched on at 37 °C, the human body temperature (Chowdhury et al., 1996). Exposure of *Staphylococcus epidermidis, S. aureus, and S. lugdunensis to* low temperature (4 °C) for extended periods resulted in alteration in pigmentation, size of the colony, and hemolytic activity (Onyango et al., 2012). The CDC Morbidity and Mortality Weekly Report (MMWR) reported that summer season (June, July, and August) shows the highest number of waterborne disease outbreaks associated with recreational water in the U.S. during 2009 – 2010 (Figure 4) (CDC MMWR, 2014).

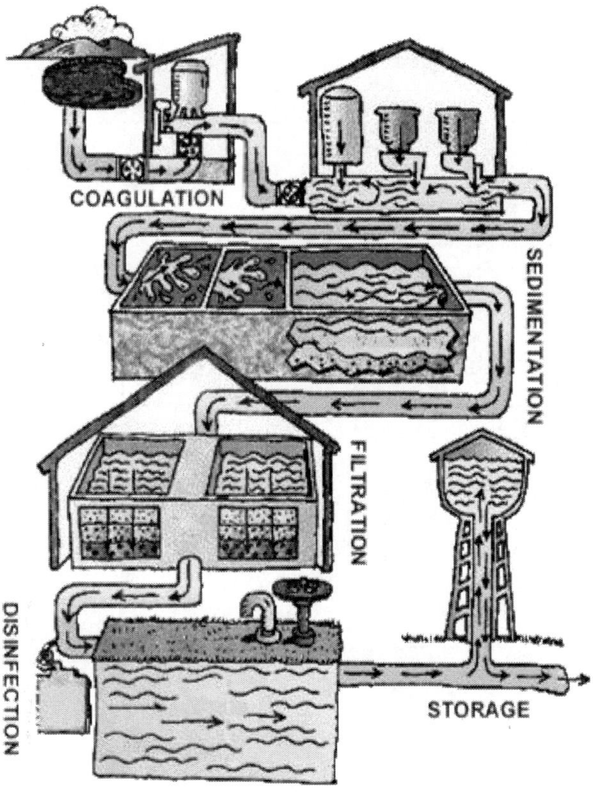

Figure 2. Water treatment process (adopted from EPA, 2014c).

Waterborne microbial pathogens are phylogenetically diverse species, encompassing eukaryotes, prokaryotes, and viruses. *Vibrio* species ranks *among the most serious* waterborne pathogens in the ocean and estuarine. *Vibrio* species are curved, motile, and facultative bacteria, some of which are symbiotic or pathogenic (Thompson et al., 2007) (Figure 5). Three major pathogenic *Vibrio* species are *V. vulnificus*, *V. parahaemolyticus*, and *V. cholerae*, which live in marine and freshwater ecosystems, are commonly found in fish and fishery products, and can cause human gastrointestinal infections (Torresi et al., 2011). It has been reported that they are responsible for foodborne infection, usually when undercooked or raw seafood is consumed. Shellfish are often contaminated with pathogenic *Vibrio* species and may cause a severe case of food poisoning. Since oysters are commonly eaten raw or partially cooked in the U.S., *Vibrio* contamination is of great concern to the oyster industry, especially for the coastal states bordered with the Gulf of Mexico - Alabama, Florida, Louisiana, Mississippi, and Texas. Besides the food poisoning problem, *V. vulnificus* can invade human body through open wounds, destructing the skin and forming ulcers and some lethal septicemia complications (Horseman & Surani, 2011).

Table 2. Waterborne outbreaks by microbial pathogens associated with recreational water in the U.S., 2009-2010 (adapted from CDC MMWR, 2014)

Etiological agent	Treated water			Untreated water		
	Outbreaks	Cases	Hospitalized	Outbreaks	Cases	Hospitalized
Bacterium						
Campylobacter jejuni	10	75	11	5	91	18
Escherichia coli	0	0	0	1	6	4
O157:H7	1	14	4	3	17	8
Legionella spp.	4	8	7	0	0	0
Pseudomonas	4	50	0	0	0	0
aeruginosa *Shigella* spp.	1	3	0	1	68	6
Parasite	**25**	**413**	**15**	**3**	**16**	**0**
Cryptosporidium spp.	24	406	14	3	16	0
Giardia intestinalis	1	7	1	0	0	0
Virus	**0**	**0**	**0**	**1**	**69**	**2**
Norovirus	0	0	0	1	69	2
Multiple	**0**	**0**	**0**	**1**	**45**	**0**
C. jejuni, norovirus genogroup I, *Shigella* sp.	0	0	0	1	45	0
Unidentified	**7**	**146**	**13**	**0**	**0**	**0**
Suspected norovirus	2	91	13	0	0	0
Suspected *P. aeruginosa*	5	55	0	0	0	0

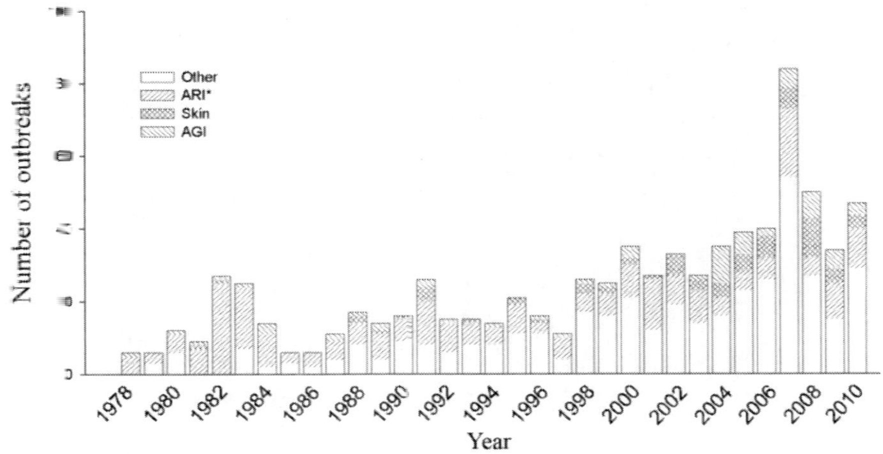

† The category of illness reported by ≥50% of ill respondents.

* All outbreaks of legionellosis (i.e., Legionnaires' disease and Pontiac fever) are classified as ARI.

Figure 3. Recreational water-associated disease outbreaks (n = 789) by predominant Illness† and Year, United States, 1978 – 2010. AGI: acute gastrointestinal illness; Skin: illness, condition, or symptom related to the skin; ARI: acute respiratory illness; Other: includes bronchitis; illness, condition, or symptom related to the ears; illness, condition, or symptom related to the eyes; hepatitis; leptospirosis; meningitis meningoencephalitis; and multiple predominant illnesses; (modified from CDC MMWR, 2014).

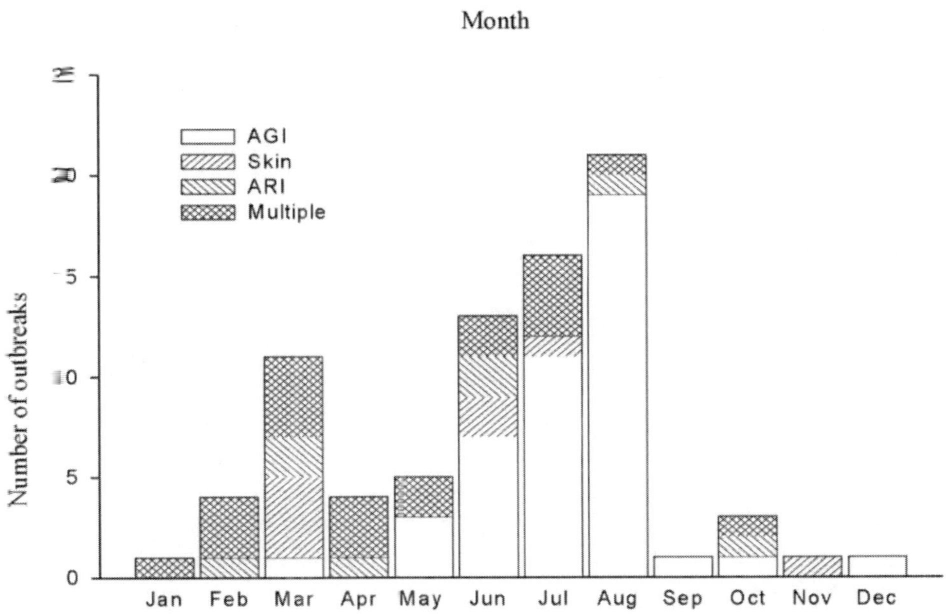

Figure 4. Number of waterborne disease outbreaks associated with recreational water (n = 81), by month in the United States, 2009 - 2010 (modified from CDC MMWR, 2014).

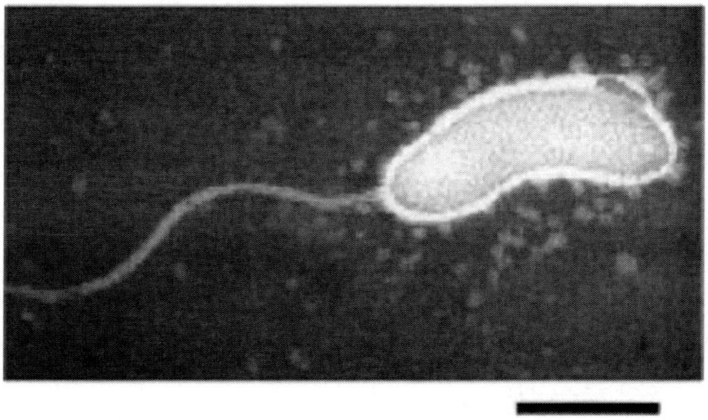

Figure 5. The planktonic *V. cholerae* visualized by transmission electron microscopy (bar =1 μm) (adopted from Watnick and Kolter, 2000).

The 2010 DH oil spill may cause significant changes in microbial community; however it is unclear whether the oil spill is beneficial or harmful to pathogenic *Vibrio* species. An abnormally high accumulation of marine pathogen *V. vulnificus* was discovered in tarballs originated from the DH oil spill (Tao et al. 2011). Tarballs, seawater, and sand samples were aseptically collected at five Gulf shore beaches in Alabama and Mississippi. Enumeration of *V. vulnificus* in these samples showed that the pathogens' abundance in tarballs is 10-times higher than that in the sand and 100-times higher than that in the seawater. While no growth was observed when the isolated *V. vulnificus* was incubated on tar-infused seawater agar for seven days. Based on that, Tao et al. concludes *Vibrio* species in marine tarballs did not necessarily degrade the oil. Instead, the *Vibrio* species may feed on by-products of hydrocarbonoclastic microorganisms in the tarballs. Another research published in 2012 and 2014 showed the presence of marine *Vibrio* species in the Gulf of Mexico after the DH oil spill by detecting ochrobactins-OH (Gauglitz et al., 2012) and amphibactin (Kem et al., 2014), siderophores of *Vibrio* species. In 2013, bacterial community in oil collected from the sea surface and sediment in the northern Gulf of Mexico showed the presence of *Vibrio* species (Liu and Liu, 2013). In their pyrosequencing study, 57 % of the total OTU (operational taxonomic units) in oil mousse collected from the leaves of *Spartina alterniflora* belonged to *Vibrio* of *Gammaproteobacteria* (Liu and Liu, 2013). Another report published in 2013 demonstrated that *Vibrio* species were present in skin and mucus of healthy Red Snapper captured in the Northern Gulf of Mexico after the DH oil spill (Arias et al. 2013). Smith et al. isolated 17 potentially-pathogenic *V. parahaemolyticus* from the Louisiana coast right after the DH oil spill (Smith et al., 2012). These *V. parahaemolyticus* isolates represent five distinct genotypes, all of which showed no oil-degrading activity. While the oil-degradation ability of *Vibrio* species, including *V. parahaemolyticus*, was previously reported (West et al., 1984; Rosano-Hemandes et al., 2012).

Historically, these *Vibrio* species and other microbial pathogens have been frequently caused waterborne disease outbreaks around the Gulf of Mexico regions, such as Lake Pontchartrain. Lake Pontchartrain is a large, shallow, brackish estuary system located north of New Orleans, Louisiana (Brammer et al., 2007). It is hydraulically connected to the Gulf of Mexico through Lake Borgne. The lake is reported to be contaminated by stormwater runoff, fertilizers, wastewater discharge from coastal treatment plants, and oil spills (Sheikh, 2005). In August 2005, Hurricane Katrina caused a huge flood in New Orleans due to the storm surge from the lake and loss of several levees (Sinigalliano et al., 2007; Nigro et al., 2011). Floodwater was pumped back into the lake after the Hurricane, which introduced gasoline, diesel, heavy metals, sewage water, pesticides, and toxic chemicals to the lake. It is estimated that more than 6.5 million gallons of oil were released into the lake during the Hurricane (Sheikh, 2005). In August 2012, Hurricane Isaac flooded New Orleans and caused the similar impacts on the lake. The type and amount of oil, physiochemical characteristics of water, and prevailing temperature determine the overall effects on microorganisms (Beazley et al. 2012). High levels of fecal coliform bacteria were observed in floodwaters of Lake Pontchartrain after Hurricane Katrina (Presley et al., 2006; Sinigalliano et al., 2007). The dominant pathogens isolated in the lake shortly after the floodwaters had receded were *Vibrio* spp., *Legionella* spp. (Sinigalliano et al., 2007), and *Aeromonas* spp. (Presley et al., 2006). From August 29 to September 11, 2005, 22 new cases of *Vibrio* illnesses were reported with five deaths among the residents of Mississippi and Louisiana. After the hurricane, the most frequently reported *Vibrio* illness was wound infections caused by *V. parahaemolyticus* and *V. vulnificus*, which represented an increase over the normal reported incidence of *Vibrio* wound infections in the Gulf coast states (CDC, 2005).

During and after the DH oil spill, more than one million gallons of two types of dispersants (Corexit® 9500A and Corexit® 9527A) were used in order to clean up the spilled crude oil in the Gulf of Mexico. Although Corexit is approved by the EPA (EPA, 2011b), many scientists claim an adverse impact on environments from its application. The dispersant shock on enteric bacteria was previously reported (For a review, see Nickerson & Aspedon, 1992) and thus it is considered that the Corexit may have certain effects on microbial pathogens especially fecal coliform bacteria. To the best of our knowledge, there is no research being conducted on this issue.

3. Analysis

3.1. PCR-Based Techniques to Detect Microbial Pathogens

Traditionally, culture-based methods had been used to detect and identify pathogenic microorganisms in environmental samples, which are labor-intensive, time-consuming, and sometimes expensive (Ranjard et al., 2000; Rompré et al., 2002; Espy et al., 2006). Presence of VBNC microorganisms also disproved the use of culture-based detection methods. For these reasons, culture-independent molecular techniques became mainstream methods to analyze microbial community structures in the environments. Figure 6 shows major strategies of culture-independent molecular techniques. All methods start with total DNA extracted from environmental samples. The total DNA is a mixture of DNAs from various micro- and macro-organisms in the samples. In order to simply detect the target microorganisms, conventional polymerase chain reaction (PCR) is used to amplify a gene that is unique in the target microorganism, such as *toxR* of *V. parahaemolyticus* and *mip* gene of *L. pneumophila* (Table 3). Amplified PCR products were then analyzed by gel electrophoresis and DNA sequencing to confirm their identity (Figure 6). Quantitative PCR (qPCR) is an advanced

PCR-based method to detect and quantify the target gene or genes (multiplex qPCR) without any further bench-work (For a review, see Smith & Osborn, 2009). In case a general microbial community needs to be monitored, the total DNA is used to amplify the gene common in group of target species, and then the mixture of PCR products are separated by subsequent molecular methods, such as denaturing gradient gel electrophoresis (DGGE), temperature gradient gel electrophoresis (TGGE), terminal restriction fragment length polymorphism (tRFLP), denaturing high performance liquid chromatography (dHPLC), amplified rDNA restriction analysis (ARDRA), single-strand conformation polymorphism (SSCP), and automated ribosomal intergenic spacer analysis (ARISA) (Widada et al., 2002). All of these methods separate the PCR products based on their G/C contents and/or nucleotides length. The separated patterns of PCR products represent the microbial community structures and DNA sequencing of each PCR products provide the identity of species. For example, bacterial 16S ribosomal RNA (rRNA) gene is widely used to analyze a bacterial community structure in the environment since the 16S rRNA gene is present in all bacterial species but their DNA sequences are generally species-specific. As an alternative method, the mixture of PCR products or total environmental DNA is directly processed by metagenomic sequencing. The massive DNA sequencing data are then analyzed by bioinformatic tools (For a review, see Kolb & Stacheter, 2013) (Figure 6).

Among culture-independent molecular methods described above, precise and rapid methods to detect the waterborne microbial pathogens are conventional PCR and qPCR targeting a unique gene in target species (Figure 6). Table 3 shows the list of genes to detect various waterborne pathogens including bacteria, viruses, and protists (Table 3). For example, Dickinson et al. (2013) detected *V. cholera*, *V. vulnificus*, and *V. parahaemolyticus* in water samples collected from Doheny State Beach, Avalon Harbor, and Catalina Island, CA using three primer sets (PVC-F_2 & PVCM-R_1, VVF & VVR, and tlhF & *tlhR).* Joly et al. (2006) identified *L. pneumophila in water samples collected from cooling towers and hot water systems in France with the primer set* mip1-A and mip2-A. Trakhna et al. (2009) identified *A. hydrophila* in a number of environmental samples, including water, plants, and fish, with *primers targeting the 16S rRNA gene of A. hydrophila* (Table 3).

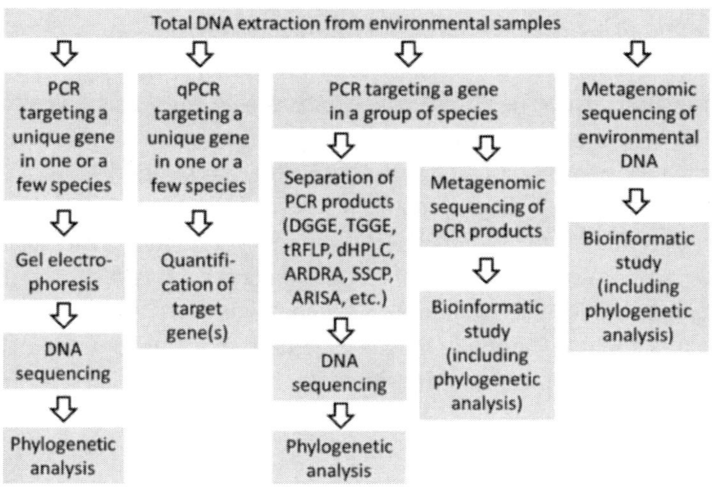

Figure 6. Major strategies of culture-independent molecular techniques.

Table 3. Primer list for waterborne pathogenic microorganisms

Primer set	Target	Reference
Bacteria		
stx1-150-F/R	Shiga toxins *stx1* gene of STEC Bacteria	Fratamico et al., 2011
stx2-200-F/R	Shiga toxins *stx2* gene of STEC Bacteria	Fratamico et al., 2011
eae-170-F/R	*eae* gene of STEC Bacteria	Fratamico et al., 2011
eco_gadA-F/R	*gadA* gene of *E. coli*	Meiland et al., 2010
eco_uspA-F/R	uspA gene of *E. coli*	Chen & Griffiths, 1998
VvAF1-F/R	16S rRNA gene of *V. vulnificus*	Vickery et al., 2007
VvBF1-F/R	16S rRNA gene of *V. vulnificus*	Vickery et al., 2007
rtxA-F/R	*rtxA* of *V. cholerae*	Chow et al., 2001
Vib_toxR-F/R	*toxR* of *V. parahaemolyticus*	Kim et al., 1999b
Vib_vvhA-F/R	*vvhA* of *V. vulnificus*	Campbell & Wright, 2003
PVC-F2/PVCM-R1	Intergenic spacer (its) of *V. cholera*	Dickinson et al., 2013
tlhF/R	Thermolabile hemolysin (*tlh*) gene of *V. parahaemolyticus*	Dickinson et al., 2013
VVF/VVR	Cytotoxin-hemolysin (*cth*)gene of *V. vulnificus*	Dickinson et al., 2013
ipaH-U1-F/R	*ipaH* of *Shigella flexneri*	Vu et al., 2004
Styinva-JHC-2-left/right	*invA* gene of *Salmonella*	Hoorfar et al., 2000
HipO-F/R	*hipO* of *Campylobacter jejuni*	Hong et al., 2007
CeuE-F/R	*hipO* of *Campylobacter coli*	Hong et al., 2007
Fe_Bactero-F/R	16S rRNA gene of *Bacteroidetes*	Dick & Field, 2004
Aero1a-F/R	*Aeromonas hydrophila*	Pollard et al., 1990
Aero_lip-F/R	*Aeromonas hydrophila* Hybridization Group 1	Cascón et al., 1996
F/R	16S rRNA gene of *Aeromonas hydrophilia*	Trakhna et al., 2009
Hp_vacA-F/R	*vacA* gene of *Helicobacter pylori*	Nayak & Rose, 2007
TUL4-435-F/R	17-kDa lipoprotein of *Francisella tularensis* LVS	Sjöstedt et al., 1997
M_rpoB-F/R	rpoB of *Mycobacterium tuberculosis*	Kim et al., 1999
mip-Lpn09-1F/R	*mip* gene of *Legionella pneumophila*	Ballard et al., 2000
LpmipFp-F/R	*mip* gene of *Legionella pneumophila*	Wilson et al., 2003
mip1-A/mip2-A	*mip* gene of *Legionella pneumophila*	Joly et al., 2006
Lep_ligA-F/R	*ligA* of pathogenic *Leptospira* spp.	Palaniappan et al., 2002
Stap_Nuc-F/R	*nuc* gene of *Staphylococcus aureus*	Brakstad et al., 1992
Y.ente-F/R	*ail* gene of *Yersinia enterocolitica*	Thoerner et al., 2003
PS23FW3/PS23RV3	23S rRNA gene of *Plesiomonas shigelloides*	González-Rey et al., 2000
Eukaryote		
Giardin P241-F/R	*Giardia lamblia* cysts	Guy et al., 2003
COWP P702-F/R	*Cryptosporidium parvum* oocysts	Guy et al., 2003
Cry9/Cry15	*Cryptosporidium* spp. oocyst wall proteins	Usluca & Aksoy, 2011
CPB-DIAGF/ R	18S rRNA gene of *Cryptosporidium*	OIE Terrestrial Manual, 2008
AWA722F/AWA1235R	18S rRNA gene of *Cryptosporidium*	Rochelle et al., 1997
Naeg-F/R	ITS of *Naegleria* spp.	Pélandakis et al., 2000
Viruses		
HAV+2030/2211	capsid protein VP3 of hepatitis A	Jansen et al., 1990
COG1F (+)/(-)	Norovirus genogroups I	.Loisy et al., 2005
QNIF2d (+)/(-)	Norovirus genogroups II	.Loisy et al., 2005
NV1LCF/R	Norovirus genogroups I Capsid	Svraka et al., 2007
NV2LCF/R	Norovirus genogroups II Capsid	Svraka et al., 2007
ABC F/R	Rotavirus VP1	Svraka et al., 2007
AC1/AC230	Astrovirus Capsid	Svraka et al., 2007
HEXAA 385/1913	Adenovirus Hexon protein	Svraka et al., 2007

3.2. U.S. Government Regulations and Risk Assessments in the Gulf of Mexico

The Federal Water Pollution Control Act of 1948 is the first comprehensive legislation to control water pollution in the U.S. (EPA, 2014b). Later, the U.S. government promulgated the Federal Water Pollution Control Amendments of 1972, the Water Quality Standards Regulation of 1975, the Clean Water Act of 1977, the Water Quality Act of 1987, and the Beaches Environmental Assessment and Coastal Health (BEACH) Act of 2000. Under the BEACH Act, states and territories were required to develop and adopt protective water quality standards for pathogens and pathogen indicators in coastal recreational waters (Wymer, 2007). Currently, each Gulf state has its own program to monitor the waterborne pathogen levels at beaches (Figure 7).

The Gulf of Mexico encompasses a 1,631 miles coastline with five U.S. states. If bays and inland waters are included, the total coastline exceeds 16,000 miles in the U.S. alone (EPA, 2014a). In order to identify and solve the common problems in the Gulf regions, the Governors of five Gulf states formed the Gulf of Mexico Alliance (GOMA) in 2004 (Gooch-Moore et al. 2011). GOMA is expected to solve the problems of the health of the coastal ecosystems and economies of the Gulf by working five states together to amplify the synergy. "Water Quality for Healthy Beaches and Seafood" is one of the GOMA's current high priority issues, which includes (1) reducing risk of exposure to disease-causing pathogens, (2) minimizing occurrence and effects of harmful algal blooms, (3) identifying sources of mercury in Gulf seafood, and (4) improving monitoring of Gulf water resources (GOMA, 2013). As a part of efforts to "reducing risk of exposure to disease-causing pathogens" and to "improving monitoring of Gulf water resources", GOMA hosted round robin events (GOMA, 2013), published a white paper (Gooch-Moore et al. 2011), and prepared the molecular marker registry (GOMA, 2013). In their "Long-term Gulfwide Monitoring" document, Multi-year monitoring of nutrient fluxes and meteorological parameters (such as humidity, barometric pressure, and temperature) were published, which greatly helps to understand the fate of pathogenic microorganisms.

Despite of all these efforts, waterborne pathogen outbreaks in the Gulf region is not decreasing. When the number of vibriosis cases by *V. vulnificus* was compared between five Gulf states and the other 45 states, more than half of *V. vulnificus* cases were from the five Gulf states during 1999 to 2011 (Figure 8). In addition, the total number of vibriosis cases in the U.S. has not been decreased for 13 years (Figure 8). Besides the waterborne pathogens themselves, there are other factors contributing to the waterborne disease outbreaks. The advancement of medicine ironically increases the number of susceptible individuals; the very young, the elderly, pregnant women, and the immunocompromised (Gerba et al., 1996). Antibiotic overuse leads to microbial pathogens gain multidrug resistance (de la Cruz & Davies, 2000; Wiedenbeck & Cohan, 2011). Infectious diseases spread faster by domestic and international travelers (Royal & McCoubre, 1989). In summary, it is obvious that public health in the Gulf region is still threatened by microbial pathogens.

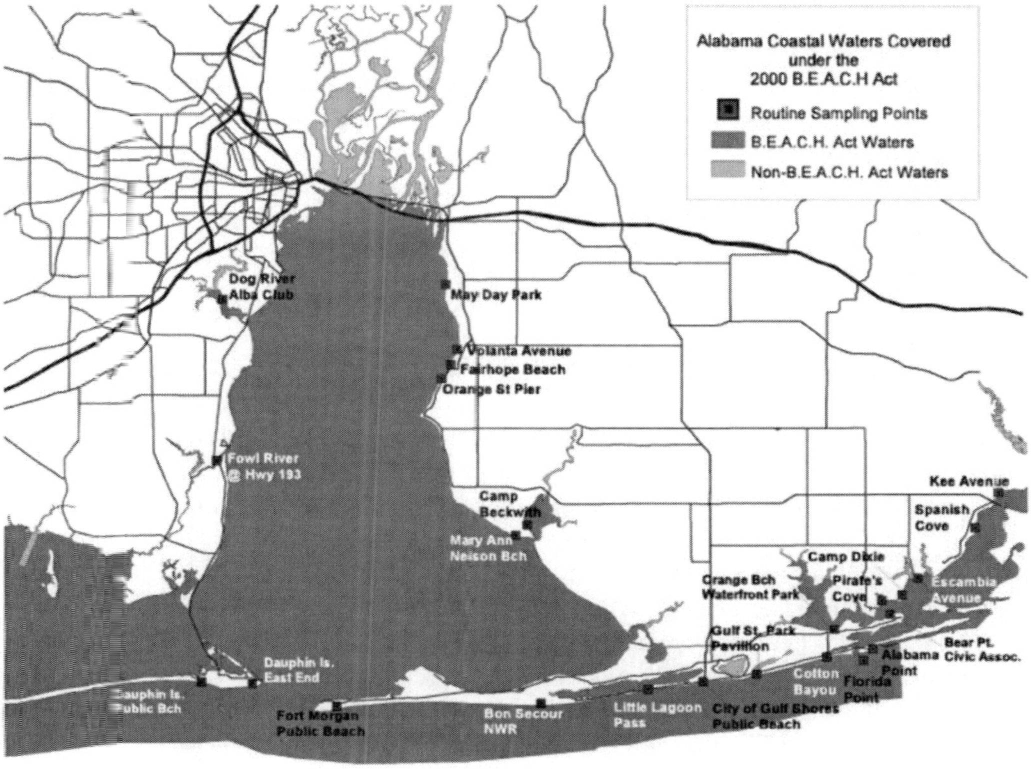

Figure 7. Alabama coastal waters covered under the 2000 BEACH Act (adapted from ADEM, 2014).

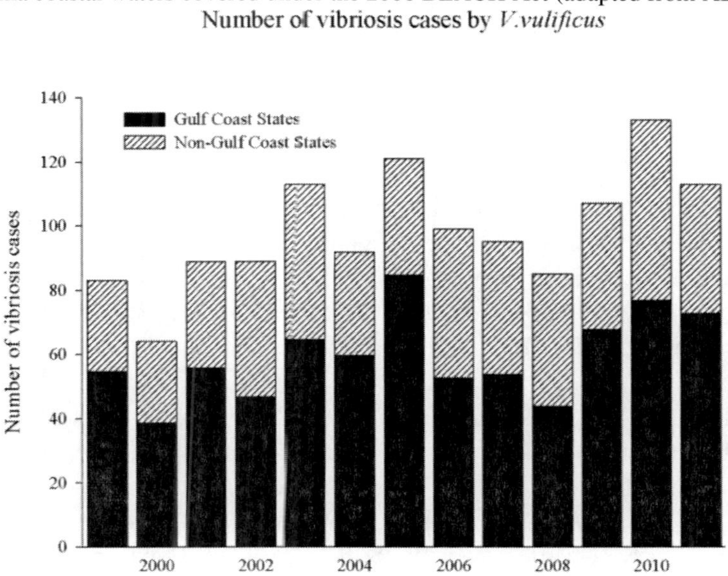

Figure 8. Number of vibriosis cases by *V. vulnificus*. (This graph is prepared based on 13 Cholera and Other Vibrio Illness (COVIS) Annual Summaries in 1999 - 2011, CDC MMWR, 2014).

There are several aspects that need to be considered to develop effective risk assessments for immediate response to potential pathogen outbreaks in the Gulf region. Basic researches should be supported by sustainable funds to develop better pathogen detection and monitoring techniques. Practical application of the research outcomes must be encouraged. Immediate response protocol to prepare future disease outbreaks can be setup based on a forecasting system in association with factors affecting the outbreak, such as temperature, salinity, current, depth, estuarine flow, hurricanes/floods, and possible anthropogenic accidents including oil spills.

CONCLUSION

This chapter briefly reviews (1) waterborne microbial pathogens and their outbreaks in the United States, (2) research progress on the microbial pathogens in the Gulf region after the DH oil spill, (3) detection strategies for major pathogens based on culture-independent molecular techniques, and (4) recent history of government policies and regulations to prevent and control waterborne disease outbreaks in the Gulf region.

Many studies have been conducted on either "waterborne pathogens" or "oil spills", but only a few studies have been conducted on the waterborne pathogens in oil-polluted environments. The Gulf of Mexico region has been experienced both the waterborne disease outbreaks and the oil spills. Moreover, foreseeable natural disasters (e.g., hurricanes and floods) and the presence of numerous oil rigs and refinery facilities around the Gulf region strongly suggest the co-occurrence of oil spills and waterborne disease outbreaks. Therefore it is crucial to integrate the information and prepare potential pathogen outbreaks following future oil spills and natural disasters in the Gulf region.

REFERENCES

Alabama Department of Environmental Management (ADEM). (2014) ADEM/ADPH Coastal Alabama Beach Monitoring Program. http://www. adem.state.al.us/programs /coastal/beachMonitoring.cnt

Arias, C. R., Koenders, K. & Larsen, A. M. (2013). Predominant bacteria associated with red snapper from the Northern Gulf of Mexico. *Journal of Aquatic Animal Health*, *25*, 281-289.

Asakura, H., Kawamoto, K., Haishima, Y., Igimi, S., Yamamoto, S. & Makino, S. I. (2008). Differential expression of the outer membrane protein W (OmpW) stress response in enterohemorrhagic *Escherichia coli* O157: H7 corresponds to the viable but non-culturable state. *Research in Microbiology*, *159*, 709-717.

Ballard, A. L., Fry, N. K., Chan, L., Surman, S. B., Lee, J. V., Harrison, T. G. & Towner, K .J. (2000). Detection of *Legionella pneumophila* using a real-time PCR hybridization assay. *Journal of Clinical Microbiology*, *38*, 4215-4218.

Beazley, M. J., Martinez, R. J., Rajan, S., Powell, J., Piceno, Y. M., Tom, L. M., et al. (2012). Microbial community analysis of a coastal salt marsh affected by the Deepwater Horizon oil spill. *PloS One*, *7*, e41305.

Besnard, V., Federighi, M., Declerq, E., Jugiau, F. & Cappelier, J. M. (2002). Environmental and physico-chemical factors induce VBNC state in *Listeria monocytogenes*. *Veterinary Research*, *33*, 359-370.

Brakstad, O. G., Aasbakk, K. & Maeland, J. A. (1992). Detection of *Staphylococcus aureus* by polymerase chain reaction amplification of the *nuc* gene. *Journal of Clinical Microbiology*, *30*, 1654-1660.

Brammer, A. J., Del Rey, Z., Spalding, E. A. & Poirrier, M. A. (2007). Effects of the 1997 Bonnet Carre Spillway Opening on infaunal macroinvertebrates in Lake Pontchartrain, Louisiana. *Journal of Coastal Research*, *23*, 1292-1303.

Campbell, M. S. & Wright, A. C. (2003). Real-time PCR analysis of *Vibrio vulnificus* from oysters. *Applied and Environmental Microbiology*, *69*, 7137-7144.

Caro, A., Got, P., Lesne, J., Binard, S. & Baleux, B. (1999). Viability and virulence of experimentally stressed nonculturable *Salmonella typhimurium*. *Applied and Environmental Microbiology*, *65*, 3229-3232.

Cascon, A., Anguita, J., Hernanz, C., Sanchez, M., Fernandez, M. & Naharro, G. (1996). Identification of *Aeromonas hydrophila* hybridization group 1 by PCR assays. *Applied and Environmental Microbiology*, *62*, 1167-1170.

Centers for Disease Control and Prevention (CDC). (2005). Vibrio Illnesses after Hurricane Katrina–multiple states, August–September 2005. *Morbidity and Mortality Weekly Report*, *54*.

Centers for Disease Control and Prevention (CDC). (2010). CDC Glossary. http://www.cdc.gov/parasites/glossary.html

Centers for Disease Control and Prevention (CDC). (2014). *Morbidity and Mortality Weekly Reports* (MMWR) of 1991 to 2014, http://www.cdc.gov/mmwr/mmwr_wk/wk_pvol.html

Chen, J. & Griffiths, M. W. (1998). PCR differentiation of *Escherichia coli* from other gram-negative bacteria using primers derived from the nucleotide sequences flanking the gene encoding the universal stress protein. *Letters in Applied Microbiology*, *27*, 369-371.

Chow, K. H., Ng, T. K., Yuen, K. Y. & Yam, W. C. (2001). Detection of RTX toxin gene in *Vibrio cholerae* by PCR. *Journal of Clinical Microbiology*, *39*, 2594-2597.

Chowdhury, R., Sahu, G. K. & Das, J. (1996). Stress response in pathogenic bacteria. *Journal of Biosciences*, *21*, 149-160.

Cook, K. L. & Bolster, C. H. (2007). Survival of *Campylobacter jejuni* and *Escherichia coli* in groundwater during prolonged starvation at low temperatures. *Journal of Applied Microbiology*, *103*, 573-583.

Craun, M. F., Craun, G. F., Calderon, R. L., Beach, M. J. (2006). Waterborne outbreaks reported in the United States. *Journal of Water and Health*, *2*, 19-30.

de la Cruz, F. & Davies, J. (2000). Horizontal gene transfer and the origin of species: lessons from bacteria. *Trends in Microbiology*, *8*, 128-133.

DePaola, A., Nordstorm, J. C., Wells, J. G. & Cook, D. W. (2003). Seasonal abundance of total pathogenic *Vibrio parahaemolyticus* in Alabama oysters. *Applied and Environmental Microbiology*, *69*, 1521-1526

Dick, L. K. & Field, K. G. (2004). Rapid estimation of numbers of fecal Bacteroidetes by use of a quantitative PCR assay for 16S rRNA genes. *Applied and Environmental Microbiology*, *70*, 5695-5697.

Dickinson, G., Lim, K. Y. & Jiang, S. C. (2013). Quantitative microbial risk assessment of pathogenic vibrios in marine recreational waters of southern California. *Applied and Environmental Microbiology*, *79*, 294-302.

Environmental Protection Agency (EPA). (2011). Drinking Water: Celebrate the Essential, http://www.cdc.gov/features/drinkingwater/

Environmental Protection Agency (EPA). (2011). *EPA science inventory*, October 15-19. http://cfpub.epa.gov/si/si_public_record_Report.cfm? dirEntryID=236773

Environmental Protection Agency (EPA). (2014). *General Facts about the Gulf of Mexico*, http://www.epa.gov/gmpo/about/facts.html

Environmental Protection Agency (EPA). (2014). *Water Quality Standards History*, http://water.epa.gov/scitech/swguidance/standards/history.cfm

Environmental Protection Agency (EPA). (2014). *Water Treatment Process*. http://water.epa.gov/learn/kids/drinkingwater/watertreatmentprocess.cfm

Espy, M. J., Uhl, J. R., Sloan, L. M., et al. (2006). Real-time PCR in clinical microbiology: applications for routine laboratory testing. *Clinical Microbiology Reviews*, *19*, 165-256.

Fratamico, P. M., Bagi, L. K., Jr Cray, W. C., Narang, N., Yan, X., Medina, M. & Liu, Y. (2011). Detection by multiplex real-time polymerase chain reaction assays and isolation of Shiga toxin-producing *Escherichia coli* serogroups O26, O45, O103, O111, O121, and O145 in ground beef. *Foodborne Pathogens and Disease*, *8*, 601-607.

Gauglitz, J. M., Zhou, H. & Butler, A. (2012). A suite of citrate-derived siderophores from a marine *Vibrio* species isolated following the Deepwater Horizon oil spill. *Journal of Inorganic Biochemistry*, *107*, 90-95.

Gerba, C. P., Rose, J. B. & Haas, C. N. (1996). Sensitive populations: Who is at greatest risk? International *Journal of Food Microbiology*, *30*, 113-123.

Ghezzi, J. I. & Steck, T. R. (1999). Induction of the viable but non-culturable condition in *Xanthomonas campestris pv. campestris* in liquid microcosms and sterile soil. *FEMS Microbiology Ecology*, *30*, 203-208.

Gonzalez-Rey, C., Svenson, S. B., Bravo, L., Rosinsky, J., Ciznar, I. & Krovacek, K. (2000). Specific detection of Plesiomonas shigelloides isolated from aquatic environments, animals and human diarrheal cases by PCR based on 23S rRNA gene. *FEMS Immunology and Medical Microbiology*, *29*, 107-113.

Gooch-Moore, J., Goodwin, K. D., Dorsey, C., et al. (2011). New USEPA water quality criteria by 2012: GOMA concerns and recommendations. *Journal of Water and Health*, *9*, 718-733.

Gulf of Mexico Alliance (GOMA). (2013). http://www.gulfofmexicoalliance.org/2013/11/round-robins/

Guy, R. A., Payment, P., Krull, U. J. & Horgen, P. A. (2003). Real-time PCR for quantification of *Giardia* and *Cryptosporidium* in environmental water samples and sewage. *Applied and Environmental Microbiology*, *69*, 5178-5185.

Hong, J., Jung, W. K., Kim, J. M., Kim, S. H., Koo, H. C., Ser, J. & Park, Y. H. (2007). Quantification and differentiation of *Campylobacter jejuni* and *Campylobacter coli* in raw chicken meats using a real-time PCR method. *Journal of Food Protection*, *70*, 2015-2022.

Hoorfar, J., Ahrens, P. & Radstrom, P. (2000). Automated 5' nuclease PCR assay for identification of *Salmonella enterica*. *Journal of Clinical Microbiology*, *38*, 3429-3435.

Horseman, M. A. & Surani, S. (2011). A comprehensive review of *Vibrio vulnificus*: an important cause of severe sepsis and skin and soft-tissue infection. *International Journal of Infectious Diseases*, *15*, 157-166.

Hurme, R. & Rhen, M. (1998). Temperature sensing in bacterial gene regulation–what it all boils down to. *Molecular Microbiology*, *30*, 1-6.

Ivnitski, D., Abdel-Hamid, I., Atanasov, P. & Wilkins, E. (1999). Biosensors for detection of pathogenic bacteria. *Biosensors and Bioelectronics*, *14*, 599-624.

Jansen, R. W., Siegl, G. & Lemon, S. M. (1990). Molecular epidemiology of human hepatitis A virus defined by an antigen-capture polymerase chain reaction method. *Proceedings of the National Academy of Sciences*, *87*, 2867-2871.

Joly, P., Falconnet, P. A., André, J., Weill, N., Reyrolle, M., Vandenesch, F., et al. (2006). Quantitative real-time *Legionella* PCR for environmental water samples: data interpretation. *Applied and Environmental Microbiology*, *72*, 2801-2808.

Kana, B. D., Gordhan, B. G., Downing, K. J., Sung, N., Vostroktunova, G., Machowski, E. E., et al. (2008). The resuscitation-promoting factors of *Mycobacterium tuberculosis* are required for virulence and resuscitation from dormancy but are collectively dispensable for growth in vitro. *Molecular Microbiology*, *67*, 672-684.

Kem, M. P., Zane, H. K., Springer, S. D., Gauglitz, J. M. & Butler, A. (2014). Amphiphilic siderophore production by oil-associating microbes. *Metallomics*. Mar 24. [Epub ahead of print]

Kim, B. J. Lee, S. H., Lyu, M. A., Kim, S. J., Bai, G. H., Chae, G. T., Kim, E. C., Cha, C. Y. & Kook, Y. H. (1999). Identification of mycobacterial species by comparative sequence analysis of the RNA polymerase gene (*rpoB*). *Journal of Clinical Microbiology*, *37*, 1714-1720.

Kim, Y. B., Okuda, J., Matsumoto, C., Takahashi, N., Hashimoto, S. & Nishibuchi, M. (1999). Identification of *Vibrio parahaemolyticus* strains at the species level by PCR targeted to the toxR gene. *Journal of Clinical Microbiology*, *37*, 1173-1177.

Kolb S. & Stacheter A. (2013). Prerequisites for amplicon pyrosequencing of microbial methanol utilizers in the environment. *Frontiers in Microbiology*, 4, 1-12.

Liu, Z. & Liu, J. (2013). Evaluating bacterial community structures in oil collected from the sea surface and sediment in the northern Gulf of Mexico after the Deepwater Horizon oil spill. *MicrobiologyOpen*, *2*, 492-504.

Loisy, F., Atmar, R. L., Guillon, P., Le Cann, P., Pommepuy, M. & Le Guyader, F. S. (2005). Real-time RT-PCR for norovirus screening in shellfish. *Journal of Virological Methods*, *123*, 1-7.

McKay, A. M. (1992). Viable but non-culturable forms of potentially pathogenic bacteria in water. *Letters in Applied Microbiology*, *14*, 129-135.

Meiland, R., Geerlings, S. E., Stolk, R. P., Hoepelman, A. I., Peeters, P. H., Coenjaerts, F. E. & Grobbee, D. E. (2010). *Escherichia coli* bacteriuria in female adults is associated with the development of hypertension. *International Journal of Infectious Diseases*, *14*, e304-307.

Mizunoe, Y., Wai, S. N., Takade, A. & Yoshida, S. (1999). Restoration of culturability of starvation-stressed and low-temperature-stressed *Escherichia coli* O157 cells by using H_2O_2-degrading compounds. *Archives of Microbiology*, *172*, 63–67.

Muoghalu, J. (2003). Priority parameters: Abiotic and biotic components. Environmental monitoring (edited by Inyang, H. and Daniels, J.). *Encyclopedia of Life Support Systems* (EOLSS), 1-8.

Nayak, A. K. & Rose, J. B. (2007). Detection of *Helicobacter pylori* in sewage and water using a new quantitative PCR method with SYBR green. *Journal of Applied Microbiology*, *103*, 1931-1941.

Nickerson, K. W. & Aspedon, A. (1992). Detergent–shock response in enteric bacteria. *Molecular microbiology*, *6*, 957-961.

Nigro, O. D., Hou, A., Vithanage, G., Fujioka, R. S. & Steward, G. F. (2011). Temporal and spatial variability in culturable pathogenic *Vibrio* spp. in Lake Pontchartrain, Louisiana, following hurricanes Katrina and Rita. *Applied and Environmental Microbiology*, *77*, 5384-5393.

OIE Terrestrial Manual. (2008). Chapter 2.9.4. - *Cryptosporidiosis*, 1192-1215.

Oliver, J. D. (2009). Recent findings on the viable but nonculturable state in pathogenic bacteria. *FEMS Microbiology Reviews*, *34*, 415-425.

Palaniappan, R. U., Chang, Y. F., Jusuf, S. S., Artiushin, S., Timoney, J. F., McDonough, S. P., Barr, S. C., Divers, T. J., Simpson, K. W., McDonough, P. L. & Mohammed, H. O. (2002). Cloning and molecular characterization of an immunogenic LigA protein of *Leptospira interrogans*. *Infection and Immunity*, *70*, 5924-5930.

Pelandakis, M., Serre, S. & Pernin, P. (2000). Analysis of the 5.8S rRNA gene and the internal transcribed spacers in *Naegleria* spp. and in *N. fowleri*. *Journal of Eukaryotic Microbiology*, *47*, 116-121.

Pollard, D. R., Johnson, W. M., Lior, H., Tyler, S. D. & Rozee, K. R. (1990). Detection of the aerolysin gene in *Aeromonas hydrophila* by the polymerase chain reaction. *Journal of Clinical Microbiology*, *28*, 2477-2481.

Presley, S. M., Rainwater, T. R., Austin, G. P., Platt, S. G., Zak, J. C., et al. (2006). Assessment of pathogens and toxicants in New Orleans, LA following Hurricane Katrina. *Environmental Science and Technology*, *40*, 468–474.

Ranjard, L., Poly, F. & Nazaret, S. (2000). Monitoring complex bacterial communities using culture-independent molecular techniques: application to soil environment. *Research in Microbiology*, *151*, 167-177.

Rochelle, P. A., De Leon, R., Stewart, M. H. & Wolfe, R. L. (1997). Comparison of primers and optimization of PCR conditions for detection of *Cryptosporidium parvum* and *Giardia lamblia* in water. *Applied and Environmental Microbiology*, *63*, 106-114.

Rompre, A., Servais, P., Baudart, J., de-Roubin, M. R. & Laurent, P. (2002). Detection and enumeration of coliforms in drinking water: current methods and emerging approaches. *Journal of Microbiological Methods*, *49*, 31-54.

Rosano-Hernández, M. C., Ramírez-Saad, H. & Fernández-Linares, L. (2012). Petroleum-influenced beach sediments of the Campeche Bank, Mexico: diversity and bacterial community structure assessment. *Journal of Environmental Management*, *95*, Suppl, S325-331.

Royal, L. & McCoubrey, I. (1989). International spread of disease by air travel. *American Family Physician*, *40*, 129-136.

Sheikh, P. A. (2005). The impact of Hurricane Katrina on biological resources. Congressional Research Service, Congressional Research Service (CRS) Report for Congress Library of Congress. Order Code RL33117.

Shiah, F. K. & Ducklow, H. W. (1994). Temperature and substrate regulation of bacterial abundance, production and specific growth rate in Chesapeake Bay, USA. *Marine Ecology Progress Series*, *103*, 297-297.

Sinigalliano, C. D., Gidley, M. L., Shibata, T., Whitman, D., Dixon, T. H., Laws, E., et al. (2007) Impacts of Hurricanes Katrina and Rita on the microbial landscape of the New Orleans area. *Proceedings of the National Academy of Sciences*, *104*, 9029-9034.

Sjostedt, A. Eriksson, U., Berglund, L. & Tarnvik, A. (1997). Detection of *Francisella tularensis* in ulcers of patients with tularemia by PCR. *Journal of Clinical Microbiology*, *35*, 1045-1048.

Smith, A. W. (2005). Tsunami in South Asia: What is the Risk of Post-disaster Infectious Disease Outbreaks? *Annals-Academy of Medicine Singapore*, *34*, 625-631.

Smith, C. J. & Osborn, A. M. (2009). Advantages and limitations of quantitative PCR (Q-PCR)-based approaches in microbial ecology. *FEMS Microbiology Ecology*, *67*, 6-20.

Svraka. S., Duizer, E., Vennema, H., de Bruin, E., van der Veer, B., Dorresteijn, B. & Koopmans, M. (2007). Etiological role of viruses in outbreaks of acute gastroenteritis in The Netherlands from 1994 through 2005. *Journal of Clinical Microbiology*, *45*, 1389-1394.

Tao, Z., Bullard, S. & Arias, C. (2011). High numbers of *Vibrio vulnificus* in tar balls collected from oiled areas of the North-Central Gulf of Mexico following the 2010 BP Deepwater Horizon Oil Spill. *EcoHealth*, *8*, 507-511.

Thoerner P., Bin Kingombe, C. I., Bogli-Stuber, K., Bissig-Choisat, B., Wassenaar, T. M., Frey, J. & Jemmi, T. (2003). PCR detection of virulence genes in *Yersinia enterocolitica* and *Yersinia pseudotuberculosis* and investigation of virulence gene distribution. *Applied and Environmental Microbiology*, *69*, 1810-1816.

Thompson, C. C., Thompson, F. L., Vicente, A. C. P. & Swings, J. (2007). Phylogenetic analysis of vibrios and related species by means of *atpA* gene sequences. *International Journal of Systematic and Evolutionary Microbiology*, *57*, 2480-2484.

Thompson, J. R., Randa, M. A., Marcelino, L. A., Tomita-Mitchell, A., Lim, E. & Polz, M. F. (2004). Diversity and dynamics of a North Atlantic coastal Vibrio community. *Applied and Environmental Microbiology*, *70*, 4103-4110.

Torresi, M., Acciari, V. A., Piano, A., Serratore, P., Prencipe, V. & Migliorati, G. (2011). Detection of *Vibrio splendidus* and related species in *Chamelea gallina* sampled in the Adriatic along the Abruzzi coastline. *Veterinaria Italiana*, *47*, 371-378

Trakhna, F., Harf-Monteil, C., Abdelnour, A., Maaroufi, A. & Gadonna-Widehem, P. (2009). Rapid *Aeromonas hydrophila* identification by TaqMan PCR assay: comparison with a phenotypic method. *Letters in Applied Microbiology*, *49*, 186-190.

U.S. Energy Information Administration. (2012). *Refinery Capacity Report*, http://www.eia.gov/todayinenergy/detail.cfm?id=7170

Usluca, S. & Aksoy, U. (2011). Detection and genotyping of *Cryptosporidium* spp. in diarrheic stools by PCR/RFLP analyses, *Turkish Journal of Medical Sciences*, *41*, 1029-1036.

Vickery, M. C., Nilsson, W. B., Strom, M. S., Nordstrom, J. L. & DePaola, A. (2007). A real-time PCR assay for the rapid determination of 16S rRNA genotype in Vibrio vulnificus. *Journal of Microbiological Methods*, *68*, 376-384.

Vu, D. T., Sethabutr, O., Von Seidlein, L., Tran, V. T., Do, G. C., Bui, T. C., Le, H. T., Lee, H., Houng, H. S., Hale, T. L., Clemens, J. D., Mason, C. & Dang, D. T. (2004) Detection of *Shigella* by a PCR assay targeting the *ipaH* gene suggests increased prevalence of shigellosis in Nha Trang, *Vietnamese Journal of Clinical Microbiology*, *42*, 2031-2035.

Watnick, P. & Kolter, R. (2000). Biofilm, city of microbes. *Journal of Bacteriology*, *182*, 2675-2679.

West, P. A., Okpokwasili, G. C., Brayton, P. R., Grimes, D. J. & Colwell, R. R. (1984). Numerical taxonomy of phenanthrene-degrading bacteria isolated from the Chesapeake Bay. *Applied and Environmental Microbiology*, *48*, 988-993.

Widada, J., Nojiri, H. & Omori, T. (2002). Recent developments in molecular techniques for identification and monitoring of xenobiotic-degrading bacteria and their catabolic genes in bioremediation. *Applied Microbiology and Biotechnology*, *60*, 45-59.

Wiedenbeck, J. & Cohan, F. M. (2011). Origins of bacterial diversity through horizontal genetic transfer and adaptation to new ecological niches. *FEMS Microbiology Reviews*, *35*, 957-976.

Wilson, D. A., Yen-Lieberman, B., Reischl, U., Gordon, S. M. & Procop, G. W. (2003). Detection of *Legionella pneumophila* by real-time PCR for the mip gene. *Journal of Clinical Microbiology*, *41*, 3327-3330.

World Health Organization (WHO). (2004). *Water Treatment and Pathogen Control: Process Efficiency in Achieving Safe Drinking Water*. Eds. LeChevallier M. W. and Au K.-K. 5-40. IWA Publishing, London, UK.

Wymer, L. J. (2007). *Statistical Framework for Recreational Water Quality Criteria and Monitoring*. Wiley-Interscience, NY.

Yaron, S. & Matthews, K. R. (2002). A reverse transcriptase-polymerase chain reaction assay for detection of viable *Escherichia coli* O157:H7: investigation of specific target genes. *Journal of Applied Microbiology*, *92*, 633-640.

Zimmerman, A. M., DePaola, A., Bowers, J. C., Krantz, J. A., Nordstrom, J. L., Johnson, C. N. & Grimes, D. J. (2007). Variability of total and pathogenic *Vibrio parahaemolyticus* densities in northern Gulf of Mexico water and oysters. *Applied and Environmental Microbiology*, *73*, 7589-7596.

In: Oil Spills
Editor: Adam Clifton

ISBN: 978-1-63321-515-3

Chapter 3

TOWARDS AN INTEGRATED OIL SPILL SYSTEM: FROM MODELLING TO THE DECISION SUPPORT TOOL

***Juan Manuel Sayol*[1,*], *Pau Balaguer*[2], *Daniel Conti*[1], *Ana Rietz*[3], *Marcos García-Sotillo*[4], *Gonzalo Simarro*[5], *Joaquín Tintoré*[1,2] *and Alejandro Orfila*[1]**

[1]IMEDEA(CSIC-UIB). Miquel Marqués, Esporles, Spain
[2]ICTS SOCIB. Ed. Naorte, Parc Bit, Palma de Mallorca, Spain
[3]Sociedad de Salvamento y Seguridad Marítima, Fruela, Madrid, Spain
[4]Puertos del Estado. Avda. del Partenón, Madrid, Spain
[5]ICM (CSIC), Pass. Barceloneta, Barcelona, Spain

Abstract

In this chapter, we present a state-of-the-art system developed to manage oil spills crises at the coastal areas of the Balearic Islands Archipelago. Specifically, we introduce a first version of an Integrated Decision Support System (IDSS) to response against oil spill emergencies which combines a numerical tool for tracking the spill (or objects) at the sea surface, i.e., the physical-geomorphological component, with a GIS tool containing the environmental and human variables. This system provides in real time the areas of probability of a spill, the potential impacts at the shore, the uses of the affected coasts and the resources available in the surrounding areas.

Keywords: Oil Spill, Numerical Modelling, Operational Oceanography, Environmental Sensitivity Index, Search and Rescue

1. Introduction

Pollution at sea as a consequence of accidental and/or illegal spills is among the largest threats to the marine environment. Spills involving oil or hazardous materials cause every

*E-mail address: jsayol@imedea.uib-csic.es

year large economical and ecological damages that, depending on the severity of the spill, its nature and the affected environment, could require decades to be recovered [1]. In the last years, the large number of incidents involving oil tankers, offshore platforms or oil pipelines resulted in an increasing concern among stakeholders and scientists on the need of developing accurate and reliable tools to forecast the evolution of those spills in the ocean [2–10]. To prevent the impact of oil spills in the environment, response plans recommend continuously monitoring the ocean and the use of forecasting systems based on Operational Oceanography to get routinely information from observations and models [11–13].

Operational systems can be composed by numerical models and/or observational platforms that provide the wind conditions and the Eulerian velocities of the ocean, which constitutes the basic information of any operational model for oil spill and Search and Rescue (SaR) operations. The evolution of the spill is then assessed using a Lagrangian model with the Operational system. Besides, the establishment of environmental sensitivity of the coasts is a key issue not only for protecting these areas in front accidental or illegal spills but also as a crucial element to make recognised the actuations that the Operational Agencies have to take in front of a potential crisis. Errors in the predicted trajectories of drifting objects are common and minimising them constitutes a very active area in Operational Oceanography where initiatives to improve both oceanic/atmospheric Operational systems and LPTAs (Lagrangian Particle Tracking Algorithms) are being undertaken [7]. The main sources of errors in oil spills and SaR operations are the result of the intrinsic complexity of the ocean dynamics and can be summarized as follows:

i) The Navier-Stokes equations governing the physics of the ocean are non-linear and therefore, due to the chaotic nature of the ocean small variations of the initial conditions will provide large deviation of the forecast fields that will increase with time.

ii) Operational models for ocean and atmospheric forecasting do not resolve the small scale but they parametrize the turbulent motions. In other words, no information at the sub-grid scale is provided.

iii) At the first stages of a real crisis the exact position of the accident is usually unknown resulting in wrong initial conditions for the forecasting model.

iv) The drift of an object on the ocean surface depends on the area exposed to wind and its angle of attack. These influences are usually included in the wind drag coefficient which is empirically determined sometimes after a trial-error process.

To mitigate to some extent these challenges, LPTA based models usually follow a stochastic approach [14, 15]: the uncertainty in the unknown parameters is quantified either in terms of a probability distribution or by any of the usual optimization procedures minimizing an error against available real ocean data (e.g., using drifters or HF-radars). However, the requirement of an on-line real time response, restricts the model presented in this chapter to spills at the ocean surface. Nevertheless, there are other approaches, that can study the three-dimensional processes including also chemical reactions, weathering, mechanical spreading and vertical movements but they are more computational expensive (e.g., [8, 9]).

The establishment of the Environmental Sensitivity (ES) of the coastline has its origins in the second half of the 70's of past century when the first classifications were conducted according to the behaviour of coastal habitats as response to hydrocarbons spills. The first examples were applied on the coasts of the USA where the Agency Responsible for oil spill mitigation, NOAA, outlined the main guidelines and criteria for elaborating sensitivity maps in order to set the Environmental Sensitivity Index (ESI). ESI is based on three major components: i) physical-geomorphological, ii) biological and iii) human use resources. These variables determine the resilience of coastal areas, understood as the ability for self-recovering after an oil spill accident. The usefulness of the ESI of a specific shoreline as a decision support tool resides in constructing hybrid forecasting systems (met-ocean models) combined together with the coastal characteristics [16].

2. Ocean Component

The dynamics of one spill at the sea surface depends on the sum of the forces acting on it. For one Lagrangian particle p, is given by the combined effect of waves, wind and currents. Neglecting viscous effects, the total velocity $\dot{\mathbf{x}}_p = \mathrm{d}\mathbf{x}_p/dt$ can be decomposed as,

$$\dot{\mathbf{x}}_p(\mathbf{x},t) = \mathbf{u}_p^{adv}(\mathbf{x},t) + \mathbf{u}_p^{dif}(\mathbf{x},t), \quad (1)$$

being $\mathbf{u}_p^{adv}$ and $\mathbf{u}_p^{dif}$ the advective and diffusive components of $\dot{\mathbf{x}}_p$. The advective component of the Eulerian velocity can be readily obtained from an ocean operational system as a combination of surface ocean currents ($\mathbf{u}_p^{curr}$), wave induced currents ($\mathbf{u}_p^{waves}$) and wind speed ($\mathbf{u}_p^{wind}$),

$$\mathbf{u}_p^{adv} = \mathbf{u}_p^{curr} + \mathbf{u}_p^{waves} + \gamma\mathbf{u}_p^{wind}, \quad (2)$$

where γ is a drag coefficient that depending on the object to track, spill type, shape, wind intensity and accuracy of ocean currents takes a value between 0.025 to 0.044 with a mean value of 0.03 [17]. The above can be applied for forecasting or diagnostic analysis as long as ocean currents, wave induced currents and wind speed are obtained from predictive or observational systems.

The wave induced currents are derived from the weakly non-linear wave propagation Stokes theory,

$$\mathbf{u}_p^{waves}(\mathbf{x}_p) = \frac{H^2\omega^2}{8c}\frac{\mathbf{k}}{k}, \quad (3)$$

where H is the wave height, $\omega = 2\pi/T$ the wave angular frequency, $c = \omega/k$ the wave celerity, $\mathbf{k}$ is the vector wave number and k its module ($k = \omega^2/g$).

Leaving aside the treatment of the diffusive component of the velocity, which will be addressed below, under this deterministic approach, the two-dimensional position of the particle at the ocean surface can be computed by integrating the velocity given by Eq. (1):

$$\mathbf{x}_p(t+\delta t) = \mathbf{x}_p(t) + \int_t^{t+\delta t} \mathbf{u}_p^{adv}(\mathbf{x}_p,t)dt + \int_t^{t+\delta t} \mathbf{u}_p^{dif}(\mathbf{x}_p,t)dt, \quad (4)$$

with $\mathbf{x}_p(t) = (lat,\ lon)$ the position at time t. The last term in the right hand side of Eq. (4) represents the diffusive component of the velocity field and is the result of turbulent processes of unresolved scales. As will be discussed in the following sections, the presented model introduces in the diffusion uncertainties derived from inaccuracies of the Operational numerical models or due to incorrect initial location.

2.1. Ocean Modelling Subsystem

As stated, the core for a LPTA is composed by an Operational system which provides a regular basis wind, waves and currents on a specific area. Many Operational Systems (both under an observational or numerical perspectives) are available around the globe such as the MyOcean MED system [19, 20]. In the current version of the system we use the Western Mediterranean OPerational model, WMOP that is a regional configuration for the Western Mediterranean Sea of the Regional Ocean Modeling System ocean model (WMOP/ROMS) although it can operate with any of the operational ocean forecasting systems available in the area such as the MyOcean IBI Ocean Forecast Service (further information on the IBI system and its products in `www.myocean.org` and on the IBI quality assessment in [21]). WMOP is operated by the Balearic Islands Coastal Observing System (`ww.socib.es`) and provides operationally ocean currents every 3 hours for a 3 days horizon [22]. The area under study covers the Balearic Sea extending from 5.8° W to 9.2° E and from 34.9° N to 44.7° N (see Figure 1). The grid has 631×539 (N_x, N_y) nodes with a resolution of $\sim 1.8\,\mathrm{km}$, allowing a good sampling of the first baroclinic Rossby radius of deformation (about $10 - 15$ km) throughout the whole area [23].

2.2. Diffusivity

Following [29–31] the diffusive term can be computed as,

$$\int_t^{t+\delta t} u_j^{dif}(\mathbf{x}, t)dt = R_j\sqrt{6D\delta t}, \quad j = 1,\, 2 \tag{5}$$

with $R_j(\mathbf{x}, t)$ a random number between in $(-1, 1)$ and $D(\mathbf{x})$ the diffusivity to be empirically determined.

The random walk term in Eq. (5) has been solved by inferring the value of the diffusivity D at the ocean surface for the time scales of interest. This term is usually taken spatially and temporally constant and different values are found for a specific regime (referred as eddy diffusivity in the literature) for an specific area [24]. [25] analysed the role of diffusion in the Western Mediterranean by measuring the error in the computation of Finite Size Lyapunov Exponents by changing the resolution of the model data. These authors found that diffusion introduces small scale irregularities on the trajectories, and also substantial dispersion at large scales. [18, 26] through the deployment of several surface drifters in a coastal region studied the spatial and temporal variability of diffusivity showing the different dispersion regimes in accordance with other authors [27, 28]: an exponential regime at small separations, a ballistic regime at a mesoscale range and a diffusive range at asymptotic distances.

Errors in the prediction of trajectories given by the LPTA will mainly be the result of wrong or poorly solved ocean fields provided by the Operational model it is reasonable

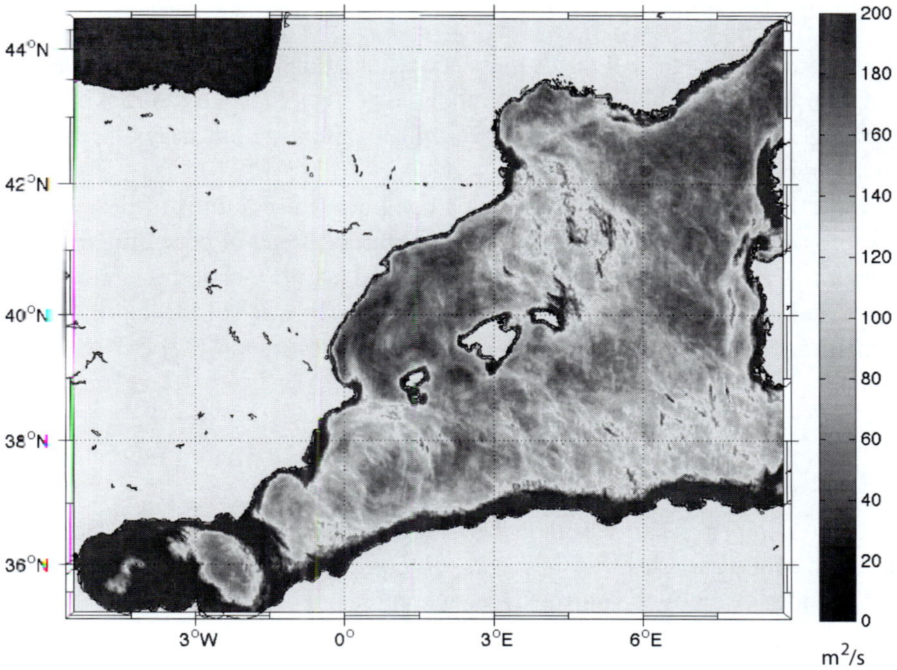

Figure 1. Geographical coverage of the numerical model. Values in the plot correspond to Western Mediterranean Sea averaged numerical diffusivity for the period January 2009-December 2011.

to compute diffusion directly from these data. Here, we have assumed the diffusivity in our model as the result of inaccuracies in the ocean model and constant in time but grid dependent. Its value is estimated directly from model data as follows. A neutrally buoyant particle was placed daily at each grid point for the period between January 2009 and December 2011 and advected during 24 hours using the velocity fields derived from the daily re-initialization and from the daily forecast. For each grid point this distance was temporally averaged and the numerical diffusivity obtained as,

$$D(lon, lat) = \frac{\epsilon^2(lon, lat)}{\delta t} \tag{6}$$

being $\delta t = 86400$ s. This pseudo-diffusivity is the result of numerical uncertainties and introduces in the LPTA the stochastic nature of the current field. For the operational model, we have temporally averaged the three year diffusivity data (see Figure 1) producing a map for the Western Mediterranean area. Larger values of diffusivity are located in the more dynamical regions (i.e., Alboran Sea, Algerian current, Liguro-Provenzal current as well as around the shelf seas) with values over $15\,\mathrm{m}^2/\mathrm{s}$. The influence of the northern winds in the Gulf of Lions and the northern Catalan coast also increase the value of the diffusivity. On average, in the Western Mediterranean $D \approx 10\,\mathrm{m}^2/\mathrm{s}$ which has also been suggested for the eddy diffusivity in other oceanic areas, with values oscillating between $1 - 200\,\mathrm{m}^2/\mathrm{s}$ [17].

2.3. Probability Domain

The chaotic nature of ocean will result in inaccuracies of the predicted fields from both the velocity data provided by the Operational model as well as those in the LPTA. Besides, at the initial stages of spill accidents the exact spill location is not always well determined resulting in large errors in predictions.

To delimit these constraints, this approach computes the contours of probability of the final positions of a group of neutrally buoyant particles instead of providing the individual track of them. A set of particles is distributed around a spill location in a circular area of small radius defined by the user. The path for each particle is solved and the probability density function of them computed by a Gaussian kernel estimator [32] as,

$$\widehat{f_{ker}}(x,y) = \frac{1}{2\pi N h_x h_y} \sum_{i=1}^{N} \prod_{j=1}^{2} \exp\left(-\frac{1}{2}\left(\frac{x_j - X_{ij}}{h_j}\right)\right), \qquad (7)$$

where N is the number of particles launched, x_j is the final position ($j = 1$ longitude and $j = 2$ latitude) of the particles, X_{ij} is the $j - th$ component of the $i - th$ observation and h_j is the bin width given by the normal reference rule, e.g.,

$$h_j = \left(\frac{1}{N}\right)^{1/6} \sigma_j, \qquad (8)$$

being σ_j the standard deviation of final positions.

With this methodology the model provides to the user contours of accumulated probability at the desired confidence intervals (e.g., 90%, 75%, 50%) which coincide with the isolines of the kernel.

2.4. Model Validation

To test the numerical component developed, a real experiment was carried out during October 2012 in the frame of the European Union project TOSCA. In collaboration with SASEMAR, Puertos del Estado and SOCIB, some drifters were deployed in the Balearic Islands region, in the middle of the Ibiza Channel. To illustrate how the system works, it has been simulated the evolution during 24 hours for one of the drifters that was deployed at $(\text{lon, lat}) = (0.8097°\,\text{E}, 38.8463°\,\text{N})$. The drifter transmitted its position through GPS from October 25th 2012 to January 23rd 2013, travelling hundreds of km until reaching the Strait of Sicily (see Figure 3(a), black line). A zoom highlighting the trajectory is displayed in Figure 3(b), where the red line shows the trajectory of the drifter corresponding to the period of simulation, from 25th October at 21:30 UTM to 26th October at 21:30 UTM. The blue and green areas represent the initial distribution of the virtual particles and the final position after 24 hours of simulation respectively. The Gaussian kernel of density probability computed from the final distribution of particles is shown in Figure 3(c) and the resulting contours of accumulated probability derived from the kernel are shown in Figure 3(d) where contours correspond to the 50% in black and 70% in grey of accumulated probability. As seen, the drifter position is inside the contour of 50% of accumulated probability (blue cross), which is also in

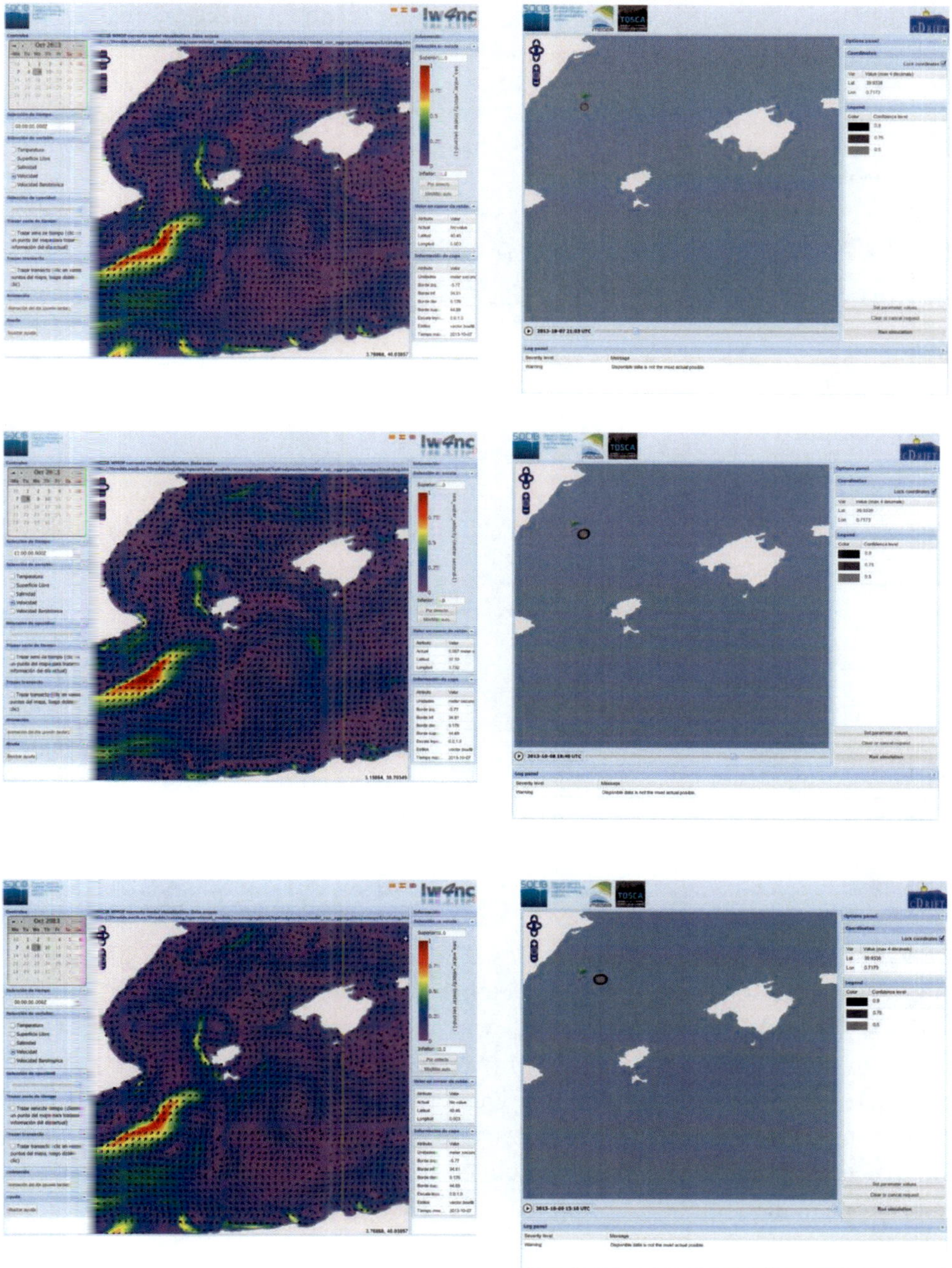

Figure 2. Daily forecast of contours of 50 (black), 70 (gray) and 90 (gray dashed) accumulated probability of spill incidence for a spill at 39.93° N and 0.72° E on October 7th 2013 (right panels). On the left the oceanographic conditions as given by the numerical model.

good agreement with previous results presented in [10]. At this point, we want to remark that several tests were made to check the proper numerical algorithm, the number of initial particles and the integration time step in order to get the best combination of accuracy and computational time. From these tests, we conclude that no differences are obtained for the final probability distribution when using more than 200 virtual particles/simulation.

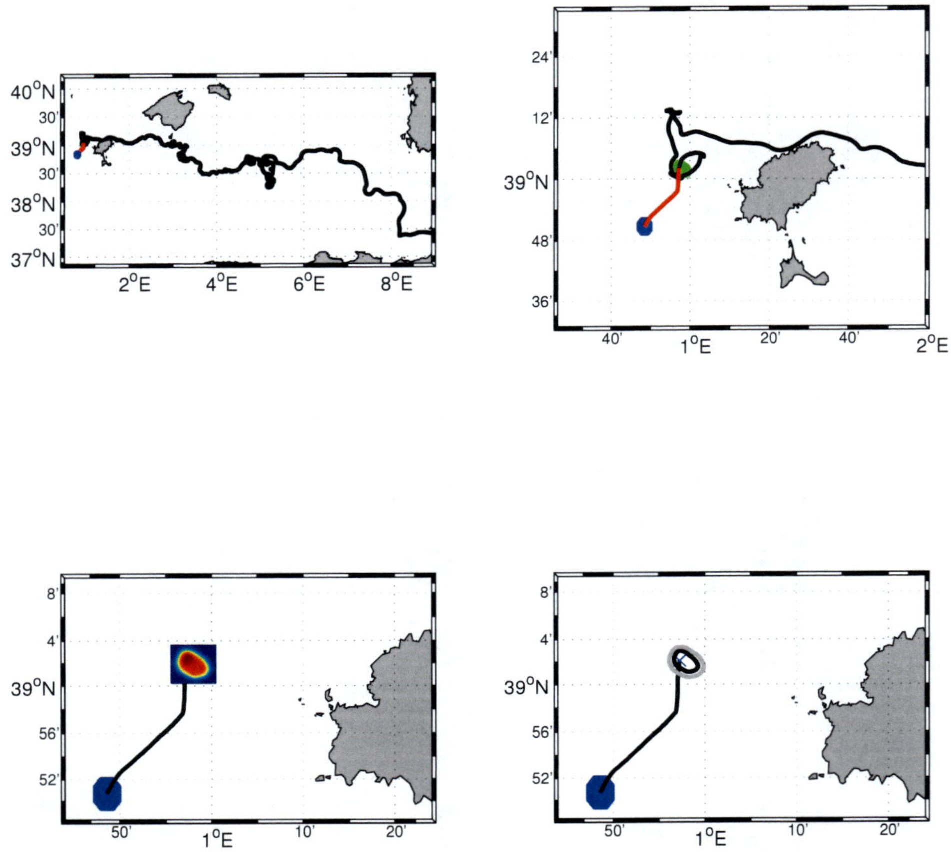

Figure 3. (a) Drifter trajectory between October 25th, 2012 to January 23rd, 2013. (b) Trajectory of the drifter for the numerical experiment corresponding to 25th October at 21:30 UTM to 26th October at 21:30 UTM (red). The initial and final distribution of the virtual particles are displayed in blue and green respectively. (c) The kernel of density probability computed for the final distribution of particles. (d) The corresponding contours of accumulated probability. The black contour corresponds to the 50% probability contour and the grey one to the 70%.

3. Decision Support System

3.1. Environmental Sensitivity Index

Shoreline Environmental Sensitivity Maps have been developed in many coastal areas of the world with examples in the most populated areas of North, South and Central America, Europe, Mediterranean Sea Region, Middle East, Indian Ocean islands, Oceania, Africa, Asia and the Pacific [33–44, 47].

The establishment of the ES of the coastline appears as a dynamic and continuous task due to the frequent changes that occur in the shoreline (variations in roughness of the coastline due to erosion or human action -i.e., ports, jetties, groynes, etc. The changes in the coastal zone may also be due to increased knowledge about these areas from a physical and/or biogeographic vision, in this way it could provide a better understanding of the sensitivity degree. The coastline may also be related with changes of spatial planning (land use changes, population increase, restoring natural areas) that will influence the pressure on these areas (e.g., new infrastructure and services). In general, knowledge of the coastline is constantly evolving and natural processes and human-induced changes makes it as a dynamic space both in the socio-economic dimension as in the natural environment (physical and biotic) both in space and time.

Integrated Coastal Zone Management (ICZM) is based on the concept by which the management of resources and terrestrials and marine areas should be fully integrated as well as the ecosystems that are part of these areas. ICZM has a close relationship with Marine Spatial Planning (MSP) defined as a practical process for establishing a rational organization of the marine environment and assessment of use. The overlapping of the study areas for IZCM and MSP gives rise the concept of Integrated Coastal and Marine Management (ICMM) [45] incorporating the marine environment of the coastal area as a fundamental element. Shoreline ES gathers information of physical, bio-ecological and human use resources constituting a decision tool to manage marine pollution emergencies, and to provide extremely useful background information in a framework of ICMM [46].

3.1.1. Shoreline Classification

Physical substrate of shoreline will determine the persistence of pollutants on the coast depending on the capacity of the materials to retain them. A description of shoreline types is shown in Figure 4. The shoreline classification is derived from 1) the height of the specific coast; 2) the exposure of the coast to the main wave climate; 3) the type of substrate; 4) the biological sensitivity degree.

- Height criteria. This criteria has been determined according to the first works related to coastal geomorphology classification for the Balearic Islands [48] which establish 3 meters height as the boundary between high and low coasts. Coastal height also will determine the accessibility for cleaning up and restoration tasks in case of a contamination event. Many rocky coasts (low and high coasts) of Balearic Islands have a step-like morphology and in some cases this feature results in a supra-tidal/backshore strip composed by accumulation of boulders, pebbles and unconsolidated materials with other sensitivity respect the materials located at the shoreline. Based on this, in recent field surveys we have adopted the criteria of [47] to map the parallel backshore strips or belts which may be affected due storm surges.

- Exposure degree to wave energy. Following [37, 49] there are mainly two types of coasts regarding the influence of waves. The first type corresponds to those coasts exposed to waves (high energy coasts). In these environments the spill will tend to be naturally removed. The second type correspond to sheltered coasts (low energy environments) where the spill will be gradually removed. The degree of exposure

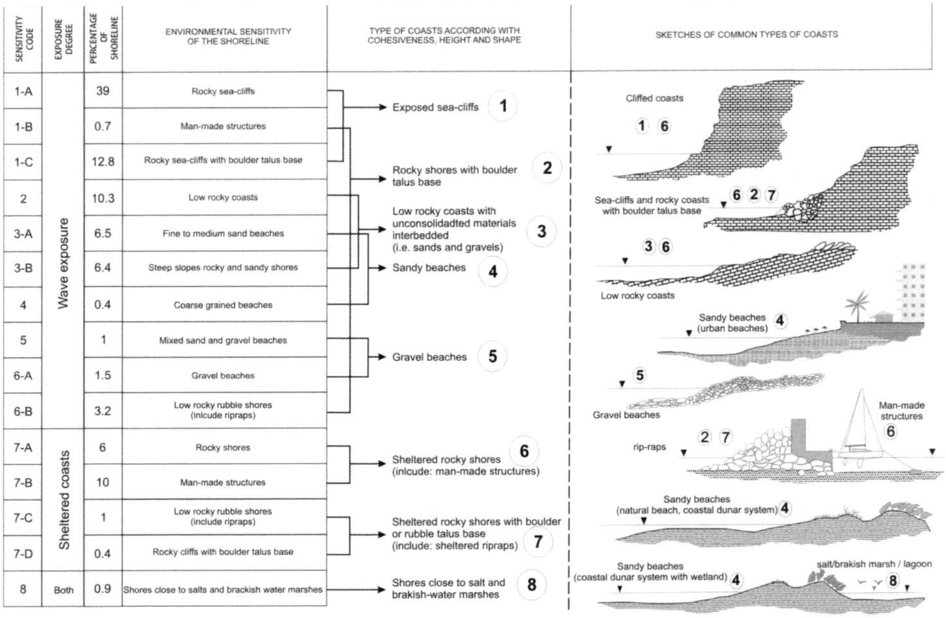

Figure 4. Description and percentage length of types of coasts of Balearic Islands according to the Environmental Sensitivity (ES). The figure represents the relationship between sensitivity degree (left), geomorphological classification (centre) and common types of coasts (right).

to wave energy is determined by the degree of articulation of the coastline (capes, natural harbour, bays, coves) and bathymetry that will condition the dissipation of the wave energy allowing the arrival of wave trains to the shoreline.

- Substrate type: the most important distinction is between bedrock and unconsolidated sediments [37]. Penetration depth is controlled by grain size of the substrate and will lead the long-term biological impacts. Deepest penetrations occur in sediments from gravel sizes ($<$ 2mm), and man-made ripraps (boulder size) that allow deeper penetration of pollutants and act as storage areas for long periods of time. Regarding the classification of beach types in relation to the grain size, it has been used previous works of textural classification for beaches of Mallorca and Menorca [50].
- Biological sensitivity. This criterion refers to the ability of the coastal environment to host different types of habitat. Indeed, coastal types are related to the morphology of the shore-in a regional context- which is not a closed concept. This classification was made upon previous works and sported by aerial photographs [48].

3.1.2. Biological-Ecological Characterization

Biological and/or ecological characterization are based on the development of different degrees of biological vulnerability of the coasts of the Balearic Islands. These degrees of vulnerability have been made according to the criteria based on a number of protection figures (marine and terrestrial) located at coastal areas in accordance with the Coastal Sensitivity Atlas of Finisterre Region [40]. In this task we have considered the kinds of protection at several levels: International (World Heritage status), European (Nature Network 2000), National and Regional (e.g., National Parks, Natural Reserves, Marine Reserves levels, among others). In this sense, natural areas are protected because of their physical, biological and ecological elements that confers them significant vulnerability. The overlapping of maps of protected areas of Balearic Islands was made using geo-processing tools (program ArcMap 9.2, ESRI).

3.1.3. Human Use Resources

The third component of the ES of the shoreline is the human use resources. This task is made by identifying the points/areas of interest in case of an oil spill or other pollution events. Features considered are infrastructures, equipments, services, leisure activities (e.g., water sports) and historical/cultural areas (including archaeological sites) of the shoreline that have to be taken into account in case of an oil spill because they could receive direct impact or could be damaged during the clean-up tasks and/or restoration of the shoreline.

3.2. The Balearic Islands Case

Shoreline Classification

The classification of the Environmental Sensitivity (ES) is made for the Balearic Islands coasts (Figure 5) being based on the international standard/guidelines proposed by NOAA designed for the establishment of the Environmental Sensitivity Index (ESI) [37]. The adoption of these guidelines responds to the need for standardizing the information in terms of the response to pollution and the location of sensitive resources that may be affected. ES classification, as a continuous task, had added more criteria from further studies published after the first results in 2006 (e.g., [47]). Classification and mapping tasks were performed through Geographic Information Systems (GIS) by using ArcGIS© program by Esri. GIS allows an agile management to integrate the information according to certain attributes during emergencies and also for coastal management purposes. ES maps comprise three types of information: a) shoreline classification; b) biological-ecological characterization; and c) human use resources.

In the Balearic Islands, we have identified 8 types of coasts divided in 15 sub-types (Figure 4). The predominant types of coasts of Balearic Islands are exposed cliffs representing the 51% of the shoreline (Figure 4, types 1-A and 1-C); coasts formed by unconsolidated materials (sandy and gravel beaches) representing the 9.4% (Figure 4, types 3-A, 4, 5 and 6-A); sheltered rocky coasts (Figure 4, types 7-A, 7-B, 7-C and 7-D) representing 17.4% being the most sensitive shorelines (Figure 4, type 8). To be note the man-made ripraps (Figure 4, included in coasts 6-B and 7-C), that are around the 2% of the Balearic shoreline located at both sheltered and exposed energy regimes. Ripraps are coasts with high secondary porosity which permits the storage and persistence of oil (or pollutants) leading to potential long-term biological impacts and making the clean-up tasks more difficult [37].

The first ES classification comprised 9 main types of coasts and 17 sub-types [51]. A reduction in the number of types of coasts was due to the suppression of sheltered sandy

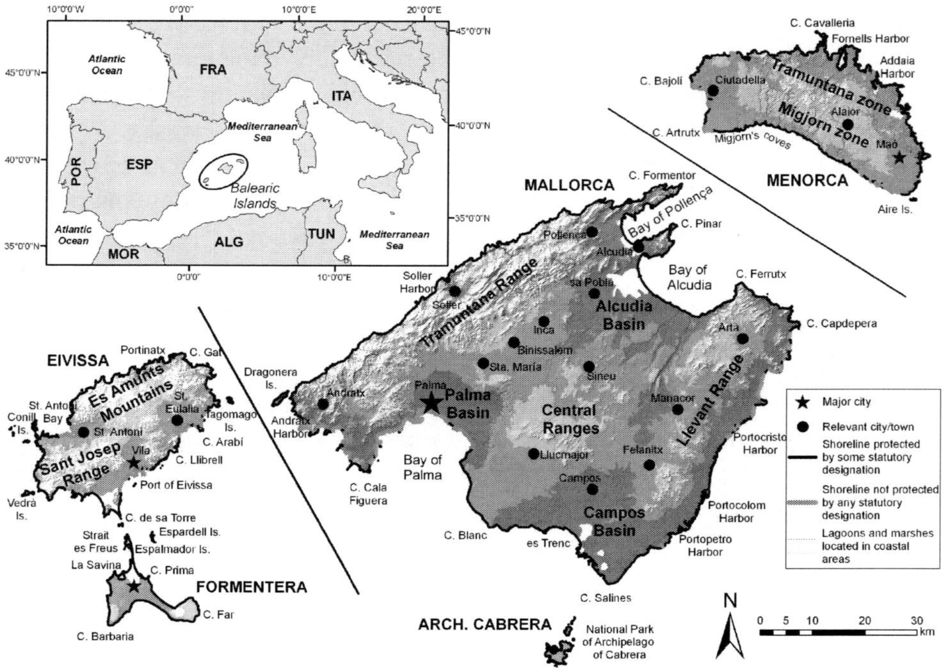

Figure 5. Location of Balearic Islands in Western Mediterranean and cartography of Balearic Islands showing main coastal toponyms, morphostructural units, cities/towns and shoreline protected by some statutory designation.

and gravel beaches. Beaches play an important role in the overall economy of the Balearic Islands hosting the major part of mass tourism [52]. The legend was changed due to: 1) the difficulty to establish the wave regime for all the beaches of the archipelago (around 900 sedimentary deposits have been recognized for the whole archipelago), b) [37] determine the sensitivity of beaches according with the capability of the substrate to sustain pedestrians and machinery for cleaning-up tasks, and c) there is not a direct relationship between the degree of exposure to wave energy and biological richness, but some studies highlights the trend of sheltered beaches to have higher biodiversity, particularly regarding macrofauna [53–55]. Therefore, exposure to beaches wave energy should be considered in respect of the coastal context, depending on the exposure to waves of nearby rocky shorelines (e.g., beaches inside caves, enclosed bays or natural harbours). During summer 2011 a field survey was carried out in the Bay of Palma (as a pilot area). This field survey campaign, along 125 km, was intended to provide better support for decision making, in connection to clean-up and restoration tasks of the shoreline. This experiment provided detailed in-situ high resolution photographs (useful for decision makers) and some features as the existence of backshore strips (accumulation of boulders and pebbles) and litoral caves were included into the cartographic database following the indications of [47].

Biological Ecological Characterization

The overlapping of protected areas in the Balearic Islands shows that 83% of the shoreline is protected by some statutory designation and the remaining 17% mainly corresponds to consolidated urban nucleus with majority of modified coasts by man and man-made structures. All the islets are protected since 1991 by a regional law, and most of them were subsequently affected by several environmental protections (e.g., Natura 2000 Network) due to their high biogeographical value because of the many endemic species of flora and fauna [56]. Shorelines with higher environmental protection are located in the strait of Es Freus, between Eivissa and Formentera, Archipelago of Cabrera (National Park) and coastal domains of Serra de Tramuntana in Mallorca (Figure 5). High environmental protection it is also noticeable in the northern coasts and central stretch of southern coasts of Menorca, northeastern and southeastern coasts of Mallorca, northwestern coasts of Eivissa and the entire shoreline of Formentera (Figure 5).

Human Use Resources

It has been mapped 27 features in the coastal zone according to their vulnerability or usefulness front an oil spill. Areas with a greater availability of human use resources are located in urban areas and especially in ports. Coastal areas with a certain lack of human use resources aimed to mitigate pollution impacts and promote environmental restoration are located in the northwestern coasts of Mallorca (Serra de Tramuntana) and Eivissa (es Amunts mountain range), and northern coasts of Menorca (Tramuntana zone) (Figure 5).

4. Results

The tools conforming the system presented in this chapter have been developed by different institutions and are continuously under upgrade. The main objective is to provide to the end-user (the agencies in charge of the response in front an accident) efficient, validated and friendly tools to react with scientific knowledge in case of a spill or SaR operations. The system has been designed to be easily relocatable as well as machine independent. Moreover, requests for a specific crisis can be made by stakeholders or civil protection personnel with little oceanographic knowledge, so the interface between model and the end-users has to be as simple as possible providing the least unknown parameters undetermined. The Operational model provides surface currents, wind and wave fields for the following 72 hours at a 3 hour interval. Velocity fields (ocean surface currents and wind speed) are arranged in a cube (tensor) with dimensions (N_x, N_y, N_t), where $N_t = 25$ corresponds to the forecasting times of $t(\text{hours}) = 0, 3, 6, \ldots.72$, provided by the WMOP.

Presently, atmospheric, oceanic and wave models (HIRLAM, WMOP and WAM respectively) are written in Fortran as well as the LPTA module, providing data in NetCDF format. After LPTA computations the resulting probability contours are displayed by GeoJSON scripts through a web based Java interface. The end-user is allowed to select some basic parameters such as the initial position (latitude and longitude), the total advection time (up to 72 hours), the number of particles to be initially deployed or the confidence intervals to be displayed. Snapshots of the Java interface for a hypothetical spill occurring at

39.93° N and 0.72° E on October 7th 2013 are shown for different times in Figure 2 (right panels). In these snapshots the specific contours correspond to the 50%, 75% and 90% of accumulated probability. The corresponding model forecasts are shown in the left panels of Figure 2.

Once the spill reaches the coast a web-based map viewer displays the cartographic data related to ES of the shoreline of the Balearic Islands (Figure 6). This viewer was designed to be a decision-making tool to support the Emergency Services of the Balearic Islands to response to potential oil spills. This Web Map Service (WMS) tool complies with OGC (Open Geospatial Consortium) interoperability standards and the criteria of the INSPIRE directive (Infrastructure for Spatial Inf). Data can be accessed directly through the map viewer, via Google Earth (metadata window and download KML/Z) or via WMS. Cartographic information is divided according to the three main components described above: 1) Shoreline types shown in Figure 4, 2) Biological resources (coastal protected areas by some statutory designation) and, 3) Human use resources. Oblique aerial photographs (corresponding to 4 different flights) are also included in order to help clean-up and recovery tasks (Figure 6). Each coastal stretch is related with an oblique aerial photograph and WMS also provides information on the length (m), description of the shoreline type and relevant information (e.g., presence of sea caves, characteristics of substrate on the backshore and variations in the legend) (Figure 6).

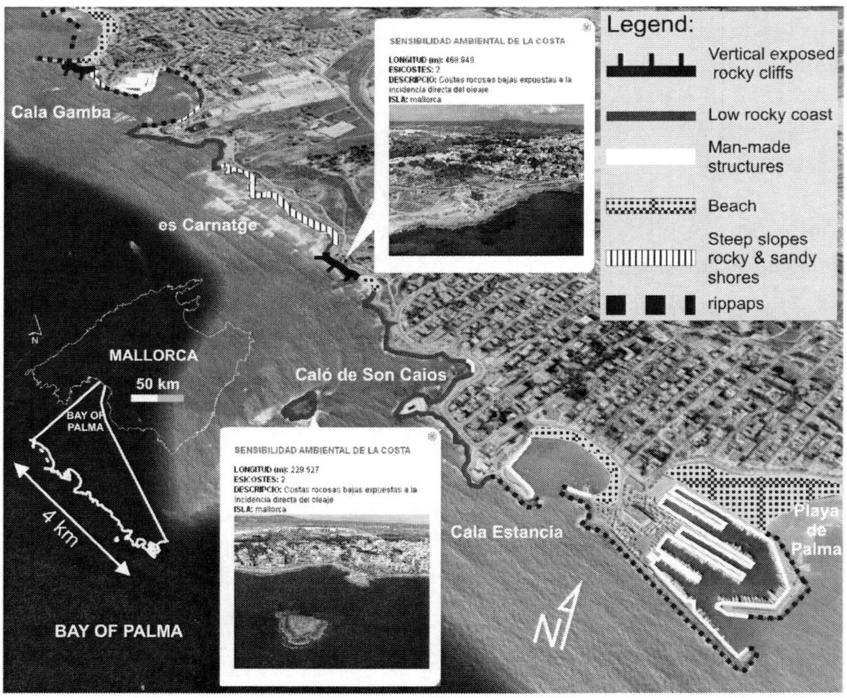

Figure 6. Example of simplified map of shoreline environmental sensitivity (ES) of a coastal stretch of Bay of Palma (Mallorca) from web-based map viewer (http://gis.socib.es/sacosta/composer).

5. Conclusion

This Chapter presents a state-of-the-art system for the management of oil spill crisis at sea. Specifically we develop a first version of an Integrated Decision Support System (IDSS) for oil spill reaction that combines a numerical tool for tracking the spill at the ocean (the physical-geomorphological component) with a GIS tool containing the environmental and human variables.

The system is conformed by several modular components that are easily relocatable and with the objective to provide to the user, accurate and updated information on the trajectory of the spill, the sensitivity of the coast where the spill will shore and the resources available through the database linked to the GIS system. The system uses met-ocean forecasts from an Ocean Operational System and provides the kernel density of the trajectories of virtual particles launched in the origin of the spill that evolve with the currents, wind and waves for the next 72 hours. When the spill reaches the coast the system uses the data from an ES system on the GIS to provide all the information about the actions that has to be taken as well as the available resources. The Lagrangian tool as well as the Environmental Sensitivity have been designed mainly for oil spills, but also could be applied to face to marine pollution accidents or SaR operations.

Acknowledgments

Authors are indebted to ICTS-SOCIB modelling, observing and data center facilities. Field and modelling support from SASEMAR is greatly acknowledged. J.M.S. is supported by the PhD CSIC-JAE program co-funded by the European Social Fund (ESF) and CSIC. D.C. thanks the financial support from the Balearic Islands Government through a predoctoral grant co-funded by the European Social Fund (ESF). G.S. is supported by the Spanish government through the Ramón y Cajal program. Authors would like to thank financial support from MED Project MEDESS-4MS. The authors also thank financial support from MICINN through Project 289 445 CGL2011-22964. Authors strongly thank the Government of the Balearic Islands and FEDER through the Grups Competitius program.

References

[1] Elliot, A., Jones, B., 2000. The need for operational forecasting during oil spill response. *Marine Pollution Bulletin* 40 (2), 110–121.

[2] Reed, M., Johansen, O., Brandvick, P., Daling, P., Lewis, A., Fiocco, R., Mackay, D., Prentki, R., 1999. Oil spill modeling towards the close of the 20th century: Overview of the State of the Art. *Spill Science & Technology Bulletin* 5 (1), 3–16.

[3] Sebastião, P., Soares, C. G., 2006. Uncertainty in predictions of oil spill trajectories in a coastal zone. *Journal of Marine Systems* 63, 257–269.

[4] Kirby, M., Law, R., 2010. Oil spill, Chemical spill, Impact assessment, Monitoring, Emergency response, Shipping accidents. *Marine Pollution Bulletin* 60 (6), 797–803.

[5] Wang, J., Shen, Y., 2010. Modeling oil spills transportation in seas based on unstructured grid, finite-volume, wave-ocean model. *Ocean Modelling* 35, 332–344.

[6] Neuparth, T., Moreira, S., Santos, M., Reis-Henriques, M., 2012. Review of oil and HNS accidental spills in Europe: Identifying major environmental monitoring gaps and drawing priorities. *Marine Pollution Bulletin* 64 (6), 1085–1095.

[7] Olascoaga, M., Haller, G., 2012. Forecasting sudden changes in environmental pollution patterns. *PNAS* 109 (13), 4738–4743.

[8] De Dominicis M., Pinardi N., Zodiatis G., Lardner R., 2013. MEDSLIK-II, a Lagrangian marine surface oil spill model for short-term forecasting - Part 1: Theory. Geosci. *Model Dev.* 6, 1851–1869.

[9] De Dominicis M., Pinardi N., Zodiatis G., Archetti R., 2013. MEDSLIK-II, a Lagrangian marine surface oil spill model for short-term forecasting - Part 2: Numerical simulations and validations. Geosci. *Model Dev.* 6, 1871–1888.

[10] Sayol J.M., Orfila A., Simarro G., Conti D., Renault L., Molcard A., 2014 A Lagrangian model for tracking surface spills and SaR operations in the ocean, *Environmental Modelling & Software* 52(2), 74–82.

[11] Kamachi, M., Fujii, Y., Zhou, X., 2002. Ocean data assimilation in the Tropical Pacific: A short survey. *Journal of Oceanography* 58 (5-7), 45–55.

[12] Siddorn, J., Allen, J., Blackford, J., 2007. Modelling the hydrodynamics and ecosystem of the North-West European continental shelf for operational oceanography. *Journal of Marine Systems* 65 (1-4), 417–429.

[13] Gonzalez, M., Ferrer, L., Uriarte, A., Urtizberea, A., Caballero, A., 2008. Operational Oceanography System applied to the Prestige oil-spillage event. *Journal of Marine Systems* 72, 178–188.

[14] Abascal, A., Castanedo, S., Medina, R., Liste, M., 2010. Analysis of the reliability of a statistical oil spill response model. *Marine Pollution Bulletin* 60 (11), 2099–2110.

[15] Minguez, R., Abascal, A., Castanedo, S., Medina, R., 2012. Stochastic Lagrangian trajectory model for drifting objects in the ocean. *Stoch Environ Res Risk Assess* 26 (8), 1081–1093.

[16] Jordi, A., Ferrer, I., Vizoso, G., Orfila, A., Basterretxea, G., Casas, B., lvarez, A., Roig, D., Garau, B., Martnez, M., Fernndez, V., Forns, A., Ruiz, M., Forns, J.J., Balaguer, P., Duarte, C.M., Rodrguez, I., lvarez, E., Onken, R., Orfila, P., Tintor, J. 2006. Scientific management of Mediterranean coastal zone: A hybrid forecasting system for oil spill and search and rescue operations. *Marine Pollution Bulletin* 53, 361–368.

[17] ASCE, 1996. State-of-the-art review of modeling transport and fate of oil spills, ASCE committee on modeling oil spills water resources engineering division. *Journal of Hydraulic Engineering* 122 (11), 594–609.

[18] De Dominicis, M., Leuzzi G., Monti P., Pinardi N., Poulain P.-M., 2012. Eddy diffusivity derived from drifter data for dispersion model applications. *Ocean Dynamics*, 62, 1381–1398,

[19] Pinardi, N., Allen, I., Demirov, E., de Mey, P., ans A. Lascaratos, G. K., Traon, P.-Y. L., Maillard, C., Manzella, G., Tziavos, C., 2003. The Mediterranean ocean Forecasting System: first phase of implementation (1998-2001). *Annales Geophysicae* 21, 3–20.

[20] Dombrowsky, E., Bertino, L., Brassington, G., Chassignet, E., Davidson, F., Hurlburt, H., Kamachi, M., Lee, T., Martin, M., Mei, S., Tonani, M., 2009. GODAE systems in Operation. *Oceanography* 22 (3), 77–92.

[21] Maraldi, C., Chanut, J., Levier, B., Reffray, G ., Ayoub, N., De Mey, P., Lyard, F., Cailleau, S., Drevillion, M., Alvarez- Fanjul, E., García-Sotillo, M., Marsaleix, P. and the Mercator Research and Development Team. NEMO on the shelf: assessment of the Iberia Biscay Ireland configuration, 2013. *Ocean Science* 9, 745–771.

[22] Tintoré, J. et al., 2013. SOCIB: The Balearic Islands Coastal Ocean Observing and Forecasting System Responding to Science, Technology and Society Needs. *Marine Technology Society Journal* 47 (1), 101–117.

[23] Send, U., Font, J., Krahmann, G., Millot, C., Rhein, M., Tintore, J., 1999. Recent advances in observing the physical oceanography of the Western Mediterranean. *Progress in Oceanography* 44, 37–64.

[24] Okubo, A., 1971. Oceanic diffusion diagrams. *Deep Sea Research* 18, 789–802.

[25] Hernandez-Carrasco, I., Lopez, C., Hernandez-Garcia, E., Turiel, A., 2012. How reliable are finite-size Lyapunov exponents for the assessment of ocean dynamics? *Ocean Modelling* 36 (4), 208–218.

[26] Mantovanelli, A., Heron, M., Heron, S., Steinberg, C., 2012. Relative dispersion of surface drifters in a barrier reef region. *Journal of Geophysical Research* 117 (C11016), 15 pp.

[27] Lacasce, J., Bower, A., 2000. Relative dispersion in the subsurface North-Atlantic. *Journal of Marine Research* 58, 863–894.

[28] Lacasce, J., Ohlmann, C., 2003. Relative dispersion at the surface of the Gulf of Mexico. *Journal of Marine Research* 61, 285–312.

[29] Ross O., Sharples, J., 2004. Recipe for 1-D Lagrangian particle tracking models in space-varying diffusivity. *Limnology and Oceanography: Methods* 2, 289–302.

[30] Marinone, S., 2006. A numerical simulation of the two- and three-dimensional Lagrangian circulation in the northern Gulf of California. *Estuarine, Coastal and Shelf Science* 68, 93–100.

[31] Marinone, S., Lavin, M., Paris-Sierra, A., 2011. A quantitative characterization of the seasonal Lagrangian circulation of the Gulf of California from a three-dimensional numerical model. *Continental Shelf Research* 31 (14), 1420–1426.

[32] Martinez, W., Martinez, A., 2002. *Computational Statistics Handbook*. Chapman and Hall. CRC, Boca Raton, Florida.

[33] Hanna, R.G.M. 1995. An approach to evaluate the application of the vulnerability index for oil spills in tropical red sea environments. *Spill Science & Technology Bulletin* 2, 171–186.

[34] Krishnan, P. 1995. Research report A geographical information system for oil spills sensitivity mapping in the Shetlands Islands (United Kingdom). *Ocean & Coastal Management* 26, 247–255.

[35] Jensen, J.R.; Halls, J.N., Michel, J.A. 1998. Systems approach to environmental sensitivity index (ESI) mapping for oil spill contingency planning and response. *Photogrammetric Engineering and Remote Sensing* 64 (10), 1003–1014.

[36] Nansingh, P., Jurawan, S. 1999. Environmental sensitivity of a tropical coastline (Trinidad, West Indies) to oil spills. *Spill Science & Technology Bulletin* 5 (2), 161–172.

[37] NOAA (National Oceanic and Atmospheric Administration). 2002. Environmental Sensivity Index Guidelines. *NOAA Technical Memorandum NOS OR&R* 11. Hazardous Materials response Division. Office of Response and Restoration. National Ocean and Atmospheric Administration & Department of Comerce United States of Amrica. 89 pp.

[38] Wieczorek, A., Dias-Brito, D., Carvarhlo-Milaneli, J.C. 2007. Mapping oil spill environmental sensitivity in Cardoso Island State Park and surroundings areas, Sao Paulo, Brazil. *Ocean & Coastal Management* 50, 872–886.

[39] Assilzadeh, H., Gao, Y., Ley, J.K.. 2009. Coastal zone mapping for oil spill emergency management. *Sea Technology* 50 (9), 33-34.

[40] Castanedo, S., Juanes, J.A., Medina, R., Puente, A., Fernández, F., Olabarrieta, M., Pombo, C. 2009. Oil spill vulnerability assessment integrating physical, biological and socio-economical aspects: Application to the Cantabrian coast (Bay of Biscay, Spain). *Journal of Environmental Management* 91 (1), 149–159.

[41] Pincinato, F. L., Riedel, P.S., Milanelli, J.C.C. 2009. Modelling and expert GIS based on knowledge to evaluate oil spill environmental sensitivity. *Ocean & Coastal Management* 52, 479–486.

[42] Bello-Smith, A., Cerasuolo, G., Perales, J.A., Anfuso, G. 2011. Environmental sensitivity maps: the northern coast of Gibraltar Strait example. *Journal of Coastal Research* SI64, 875–879.

[43] Fattal, P., Maanan, M., Tillier, I., Rollo, N., Robin, M., Pottier, P. 2010. Coastal vulnerability to oil spill pollution: the case of Nourmoutier Island (France). *Journal of Coastal Research* 26 (5), 879–887.

[44] Oyepedo, J.A., Adeofun, C.O. 2011. Environmental sensitivity index mapping of Lagos shoreline. *Global Nest Journal* 13 (3), 277-287.

[45] Cicin-Sain, B., Belfiore, S. 2005. Linking marine protected areas to integrated coastal and ocean management: A review of theory and practice. *Ocean & Coastal Management* 46 (11-12), 847-868.

[46] Belfiore, S., Cicin-Sain, B., Ehler, C., Editors. 2004. *Incorporating Marine Protected Areas into Integrated Coastal and Ocean Management: Principles and Guidelines.* IUCN, Gland, Switzerland and Cambridge, UK. 38 pp.

[47] Adler, E., Inbar, M. 2007. Shoreline sensitivity to oil spills, the Mediterranean coast of Israel: assessment and analysis. *Ocean & Coastal Management* 50, 24-34.

[48] Butzer, K.W. 1962. Coastal geomorphology of Majorca. Annals of Asoc. *American Geographers* 52 (2), 191-212.

[49] Hayes, M.O. 1996. An exposure index for oiled shorelines. *Spill Science & Technology Bulletin* 3, 139-147.

[50] Gómez-Pujol, L., Roig-Munar, F.X., Forns, J.J., Balaguer, P., Mateu, J. 2012. Provenance-related characteristics of beach sediments around the island of Menorca, Balearic Islands (western Mediterranean). *Geo-Mar Lett* 33, 195–208.

[51] Balaguer, P., Sardá, R., Ruiz, M., Diedrich, A., Vizoso, G., Tintoré, J. 2008. A proposal for boundary delimitation for integrated coastal zone management initiatives. *Ocean & Coastal Management* 51, 806–814.

[52] Aguiló, E., Alegre, J., Sard, M. 2005. The persistence of the sun and sand tourism model. *Tourism Management* 26 (2), 219-231.

[53] McLachlan, A. 1990. Dissipative beaches and macofauna communities on exponed intertidal sands. *Journal of Coastal Research* 6 (1), 57-61.

[54] McLachlan, A., Brown, A.C. 2006. *The ecology of sandy shores.* 2on edition. Academic Press, New York, 373 pp.

[55] Sevastou, K., Lampadariou, N., Eleftheriou, A. 2011. Meiobenthic diversity in space and time: The case of harpacticoid copepods in two Mediterranean microtidal sandy beaches. *Journal of Sea Research* 6 (3), 205–214.

[56] Palmer, M., Pons, G., Cambefort, I., Alcover, J.A. 1999. Historical processes and environmental factors as determinants of inter-island differences in endemic faunas: the case of the Balearic Islands. *Journal of Biogeography* 26, 813–823.

In: Oil Spills
Editor: Adam Clifton

ISBN: 978-1-63321-515-3

Chapter 4

OIL SPILLS: ENVIRONMENTAL CONSEQUENCES AND RECOVERY STRATEGIES

Cristiane Sueli Talhiaferro de Araújo, Márcia Miyuki Hoshina, Maria Tereza Pamplona Silva, Matheus Mantuanelli Roberto, Raquel Vaz Hara, and Maria Aparecida Marin-Morales
Institute of Biosciences, Departament of Biology, UNESP –
Universidade Estadual Paulista, Rio Claro – SP, Brazil

ABSTRACT

Petroleum perhaps is one of the most important substances to the modern society, due to its wide use in the most diverse productive sectors. Besides serving as raw material to the production of plastics and other compounds, it is the fuel source used in the production of electric energy, in industries and in the different transport sectors. However, the ecosystems located in areas of extractions, processing and distribution of petroleum end up being highly susceptible to the direct and indirect impacts of these activities, as well as those located in their transport route. As the large extraction fields are located in coastal areas, these environments are, generally, the most affected, although other ecosystems also suffer influence of the petroleum activities. Oils spills and leakages provoke immediate effects into the environment, easily viewed by the generalized mortality of animals and consequent ecosystems losses, but also provoke more persistent effects, such as alteration of the animals behaviour and permanence of compounds derived from petroleum in the food webs. Accidents with petroleum are unpredictable and the resources needed to minimize their impacts, often, are not liable to an immediate application. Among the strategies developed to recover the impacted environments, there are the mechanical cleaning methods, use of chemical dispersant products and use of bioremediation treatments. Cleaning methods of areas contaminated by petroleum are only considered effective when they prevent the migration of this pollutant to more sensitive ecosystems. Now, the bioremediation has been considered a "green" alternative for treatment of dangerous contaminants since it uses organisms naturally able to reduce, remove and transform the organic and inorganic pollutants. The success of this process depends on it natural biodegradability, bioavailability and optimization of the biological activities related to the pollutant. Among the strategies available to perform the natural bioremediation there are the natural attenuation, bioventilation, biomagnification, biostimlation, landfarming, composting and phytoremediation. In this chapter it will be discussed the concepts related to the impacts caused by petroleum and its derivatives, as well as the ecological consequences of some petroleum accidents and the efficiency of the application of the environmental decontamination techniques, such as bioremediation, since this is an effective strategy, economically viable and that provides a lower environmental impact.

Keywords: Accidents and environmental impacts, environmental recovery, natural attenuation, bioremediation, biostimulation, biomagnification, phytoremediation

INTRODUCTION

Petroleum is one of the most important substances consumed in our society (Speight, 2014), considered as one of the most important sources of energy and raw material for the production of synthetic polymers and several chemical substances. Transportation and global use of petroleum and petroleum products made petroleum hydrocarbons the major contaminants, both by the quantity and prevalence in the environment (Tyagi et al., 2011). Thus, every time that petroleum is exploited, transported or stored, or when petroleum products are used, there is a risk of occurring spills, and, consequently, of causing significative impacts in the environment (Annunciado et al., 2005).

Spills of oil and petroleum products can occur by accidents with oil tankers (Torrey Canyon in 1967; Amoco Cadiz in 1978; Exxon Valdez in 1989; Braer in 1993; Nakhodka in 1997; Jessica in 2001), equipment failures (Deepwater Horizon - British Petroleum, Gulf of Mexico, in 2010), pipelines disruptions (OSBAT - Guaecá, São Paulo/BR, in 2004; Guanabara Bay, Duque de Caxias/RJ, Brazil, in 2000; Barigui and Iguaçu Rivers, Paraná, Brazil, in 2000), wars (Gulf War, Kuwait, 1991), besides natural events (Kocaeli earthquake, Turkey, 1999; Katrina, Rita and Wilma hurricanes, 2005) and slow leaks of natural petroleum reservoirs. These events are able to contaminate water, soil and air (Trindade et al., 2005).

In the period of 1970 to 2004, 73 accidents with petroleum were recorded, whose amount of oil involved was greater than 700t/accident. Since 1970, the number of spills considered large ($\geq$ 10,000 t) and very large ($\geq$ 100,000 t) is decreasing significantly (Burgherr, 2007). For example, in the decades of 80's and 90's, the total number of spillages decreased, in relation to the 70's decade, mainly by the reduction of large spills (from 10.000t to 100.000t).

Regarding the larger spills or equal to 700t, onshore and offshore, in the same period (1970 - 2004), the mean of the results of the second half of the 70's decade present higher amount of offshore accidents, resulted from few accidents, but of greater quantities of oil discharged into the environment. In 1980 and 1990, there was a significant decrease in these events, although the mean value of petroleum spilled onshore more than doubled in the 90's decade, mainly due to 4 accidents of very large-sized accidents (with more than 100,000 t of oil spilled/accident).The data analysis shows a trend for a decrease in the offshore accidents and an increase in the onshore spills (Burgherr, 2007).

According to González et al. (2006), dispersion and dissolution are the first processes that occur when the oil is spilled. The persistence of the petroleum depends on the quantity and quality of the hydrocarbons mixture and also of the characteristics of the affected environment (Atlas, 1991). Now, the extension of these impacts is determined by several factors such as the amount and type of oil spilled; geographical location as well as political and legal aspects of the territory; proximity to more sensitive environments; choice and efficiency of the contaminants removal strategies (McCay et al., 2004).

To prevent the extent of environmental damage and develop efficient strategies to recover an impacted environment, there is the need to perform a detailed investigation on the fate of the spilled oil (González et al., 2006).

Currently, the range of existing methodologies to remove oil and derivatives is vast and crescent. In general, these methodologies available to remove oil from the environment can be categorized in: physical, chemical and bioremediation methods (Zhu et al., 2013).

Thus, this chapter aimed to provide a short panorama of the main accidents with petroleum in different environments (soil, groundwater, inland waters, ocean waters) and describe the impacts of this contamination on living organisms, as well as elucidate the most used remediation methods to recover the impacted environments.

1. CONTAMINATIONS

1.1. Ocean Waters

The annual estimated amount of petroleum spilled on the ocean surface is between 10,000,000 and 20,000,000 t, and it is known that only 1 ton of petroleum is able to cover about 12 km^2 of the ocean surface (Cardoso, 2007).

According to Kingston (2002), when the petroleum is spilled into the sea, it initially spreads on the surface, as a spot with few millimetres of thickness, where the volatile compounds quite toxic, quickly evaporate. Still according to Kingston (2002), the lighter the oil the faster is its evaporation and, therefore, its removal from the sea surface is easier.

In the literature there are reports of oil spills that receive more attention, not only because they are considered great catastrophies but also due to the effects that are observed after several years after their occurrence, mainly in the affected populations. Until 2013, 38 accidents involving super-tankers were recorded worldwide (Janjuaet al., 2013). Gundlach and collaborators (1983) described the oil spill from the tanker *Amoco Cadiz*, occurred in March 1978, which resulted in the leakage of approximately 223,000 t of oil.

These authors estimate that about 10 thousand tons of oil was degraded microbiologically, although it was registered by other authors (Atlas et al., 1981) that, after 3 years of the accident, despite the easing of the main effects, the concentrations of hydrocarbons were still high.

In March 1989, the tanker *Exxon Valdez* caused the largest oil spill in the US history, until that moment. The spillage resulted in the leakage of 42 millions of litres of raw petroleum, which caused an oil spot that spread for approximately 3,000 km^2 and for over 350 km of preserved beaches of the region. The type of oil spilled, associated with strong winds of the region and the tidal regime, prevented the use of dispersants and the oil spot ended up being carried to the coast (Atlas; Hazen, 2011).

In January 1993, an accident with the tanker *Braer* spilled, approximately, 84 thousand tons of crude oil and 1,500 tons of fuel oil in the sea of the Shetland Islands, United

Kingdom. As the oil spot was located under the surface level, there was no wind action on it. However, the oil was spread by sea currents in the north of the contamination site, reaching regions of the western part of the islands and the southeast of the North Sea.The measurements of the oil concentration in the sea, during the period of the accident, showed quantities of oil 20,000 times higher (50 ppm) near the shipwreck and 500 times higher (1,500 ppb) in open sea, when compared to the concentrations of the areas not affected by the accident. Despite the low mortality of seabirds and local mammals, there was a prohibition of collecting fish and shellfishes, as well as their commercialization for human consumption. This accident was considered as "the worst British environmental disaster" (Kingston, 1999).

In 1996, the *Sea Empress* tanker, which carried more than 130,000 tons of light crude oil, ran aground at the entrance of the port of Milford Haven – Southwest of Wales. About 72,000 tons of crude petroleum and 360 tons of heavy fuel oil leaked into the sea and contaminated around 200 km of the coastline, which caused strong odours and consequent complaints of residents of the coastal cities, with symptoms related to this exposure (Harvey et al., 1999; Lyons et al., 1999).

Another major oil spill occurred in January 1997, when the Russian tanker Nakhodka, which carried 19 thousand tons of oil to Russia, cracked at northeast of the Oki islands in the Sea of Japan and spilled more than 6,000 tons of petroleum into the sea. The accident occurred near the coast of Anto district, in the city of Mikuni, Fukui, and the oil reached the west coast of the main island of Honshu, in Japan (Morita et al., 1999).

On December 12, 1999, the Maltese tanker *Erika* wrecked releasing in south coast of Brittany, France 5.8 million gallons of heavy oil. This accident resulted in a major leakage that impacted the aquatic life, birds, marine fauna and flora. Besides the direct lethal effects of this accident, indirect effects resulted from the contamination of the food chain, are also associated to it (LeHir; Hily, 2002; Lemiere et al., 2005).

In 2000, there was the disruption of a duct that caused a leakage in the region of Duque de Caxias Refinery, in Rio de Janeiro, Brazil, which resulted in a large-sized spill, with more than 1.3 million litres of heavy oil, visible along the 5 km of the Guanabara Bay coast. This episode occurred in a period of syzygy tide, which resulted in an impact in the mangrove areas of the Guanabara Bay and more internal areas (Maciel-Souza, 2006).

Also in 2000, the tanker *Eurobulker* sank in Greece, spilling 700 tons of crude petroleum, which reached beaches and a great portion of the coast. The attempt to recover the impacted area was done by cleaning the coastal zone, mainly by mechanic techniques (Zenetos et al., 2004).

Doval et al. (2006) and González et al. (2006) monitored the presence of aromatic hydrocarbons dispersed and dissolved in the water column of the Galicia coast (Spain), impacted by the wreckage of the tanker *Prestige*, occurred in 2002, which transported about 77,000 tons of oil. In this accident, most part of the National Park of the Atlantic Islands was seriously contaminated. According to the Technical Report CSIC-Prestige-14 (2003), the type of crude oil that was spilled in this region is considered persistent and, even after one month of the accident, there were no consistent signals of biodegradation and only 5% of its total components was lost by evaporation.

The Greek tanker *Tasman Spirit* that, in August 2003, transported 67 thousand tons of crude oil from Iran to Pakistan ran aground before entering in the channel of the port of Karachi, in Pakistan. After the cracking of the ship, there was the release of its content into the sea. The marine conditions (choppy sea and strong winds) enabled the petroleum propagation for approximately 10 km of the coast. Two more spills, in the same month, also occurred, resulting in a total leakage of more than 35,000 tons of crude oil. The part of the coast most affected was occupied by a residential and leisure area densely populated (Janjua et al., 2013).

In December 2007, the tanker *Hebei Spirit* collided with a South Korean freighter in the waters of the west sea of South Korea, releasing about 10,900 tons of raw petroleum. Due to the strong winds, the oil spill was taken towards to the coast and reached more than 375 km of beaches and rocks (Yim et al., 2012; Kim et al., 2013). According to Yim and collaborators (2012), emergency operations were performed to remove the great amount of oil until January 2008 and, after this period, secondary operations were carried out until October 2008. According to Kim et al. (2010), the oil concentrations in these waters, 15 days after the spillage, were above 16,600 g/L and, after 10 months, decreased to 10 g/L in most places. According to Yim et al. (2012) this was the greatest disaster before *Deepwater Horizon*, which occurred in the Gulf of Mexico in 2010.

Shultz et al. (2014) described the impacts caused by one of the largest oil spills ever recorded. *Deepwater* Horizon was a semi-submersible oil tower located in the Gulf of Mexico that exploded in April 2010, and resulted in the spillage of 4.9 million barrels of petroleum which reached approximately 68 thousand m^2. This disaster caused a large scale environmental contamination, which affected the entire ecosystem, besides inducing damages in the human health (Hayworth; Clement, 2011; Hemmer et al., 2011). The contamination in this region was further enhanced with the use of 1.8 million litres of dispersants, used to solubilise the oil in small micelles and, thus, facilitate the bioremediation (Hazen et al., 2010; Paul et al., 2013). The use of chemical dispersants can increase the toxicity of saturated hydrocarbons and PAH fractions (Cotou et al., 2001), as well as enhance the toxicity of the crude petroleum, which persists even after the treatment (Valentine et al., 2010). According to Kostka et al. (2011) a large amount of the oil reached the shoreline and, despite the efforts to clean the environment, a considerable part still remained impacting the coastal ecosystems.

A total of 4,835 incidents of oils spills were registered between the years of 1976 and 1996, which resulted in the spillage of 1,896,960 barrels of oil into the environment. According to Nwilo and Badejo (2004), the large amount of spills is mainly due to the following factors: corrosion of tanks or pipes (50%); sabotage (28%); activities related to the production and processing of petroleum (21%); and inability to control leakages of the oil wells, mechanical failures and inadequate procedures during loading and unloading the oil (1%).

1.2. Inland Waters

Hydric reservoirs proper for human consumption such as rivers and lakes have been increasingly target of anthropic activities that pollute the aquatic ecosystems. Due to this situation, it becomes essential the rational use and development of strategies that minimize their degradation (Lindén; Pålsson, 2013).

Among the activities that cause impact in the aquatic ecosystem, the ones of the petroleum industry stand out since they are increasingly essential to the technological and economic development in the world. As the refining process uses a large amount of water, it generates a lot of waste, which ends up being destined to rivers and lakes near the industries (Lopes, 2010).

Another important factor to be considered is regarding the petroleum impact observed in rivers, open waters and coastal waterways. These spills range from small proportion impairments, derived from continuous terrestrial leakages and illegal extraction of petroleum in the oceans, to greater impairments such as major accidents. Thus, it is necessary the precise identification of hydrocarbons released into the environment in order to perform a correct assessment and application of remediation systems in the contaminated ecosystems, besides enabling a prediction, long-term, of the triggered impacts (Wang et al., 2004).

In 1984, the tanker *Mobiloil* ran aground, releasing 170,000 gallons of residual and industrial oil of high density in the Columbia River, in the state of Oregon, United States. This leak caused the contamination of waters downstream the accident site, jeopardizing the local biota and polluting waters near the river mouth and affected local beaches in the Pacific Ocean (Krahn et al., 1986). These authors used white sturgeons as bioindicators by assays developed with bile substances, where they identified the presence of metabolites of petroleum hydrocarbons at higher levels when compared to other studies performed in areas known to be contaminated by petroleum.

In 2000, in the US, there was the leakage of 140,000 gallons of fuel oil in the tributary Swanson of the Patuxent River, affecting all water bodies downstream. Next to Chalk Point power plant, in Maryland, there was a rupture of a duct that resulted in the contamination of a swamp of 76 acres (Michel et al., 2009). Correlating the data collected in 2007 and 2009, these same authors studied the impacts of this accident, mainly on plants that dominated this brackish environment. They observed that the impact on the flora varied and some species presented significative alterations in their abundance and height, but not in their subterranean biomass, while for other species the opposite was observed and the subterranean biomass was the most affected, representing a state of stress to which the plant was subjected. Moreover, these authors tested the toxicity of water samples in amphipods, observing positive damage results in 46% of the animals.

In 2002, thousands of litres of petroleum products were spilled into the Rouge River, reaching Detroit River at 3 km downstream and then culminating in oil spots in the Lake Erie, in the region of the Great Canadian Lakes. This leak eventually impacted, approximately, 43 km of the US and Canada coast (Wang et al., 2004). The studies performed by Wang et al. (2004) showed that most part of the oil spilled was composed by lubricating oil mixed with a small portion of diesel. However, in this case, diesel had already undergone some degradation due to the local weathering associated with evaporation.

Due to the risk offered by oil spills, major oil centres have become the focus of studies for the risk assessment of the environment where they are located. The Lower Mississippi River has a region of intense naval traffic, next to the Port of New Orleans, where there are still, approximately, nine petroleum refineries (Danchuk; Willson, 2010). These same authors report that, between 2007 and 2010, 4 leakages of oil pipelines and refineries occurred, when 900 to 11.000 barrels of oil were spilled in the region. Due to these accidents and to the great imminent risk, the authors studied the river characteristics, throughout its course, until the Gulf of Mexico and developed models and materials that helped in the understanding of the environment, if another possible major spill happens, in an attempt to predict which sites would be the most likely to be impacted.

In 2004, in Brazil, there was a rupture of an oil pipeline with a flow rate of approximately 2000 m^3 of a mixture of oils of the Marlin and Espírito Santo types. This pipeline was within a conservation unity, in the region of Guaecá and São Sebastião, where there are remnants of the Atlantic Forest. According to the technical report of CETESB (2005), the Guaecá River (river with excellent water quality), the Guaecá beach and the groundwater of the region were contaminated, which severely damaged the local biota. Leme et al. (2008) showed that the waters collected in the Guaecá River, even one year after the spillage, still had petroleum hydrocarbons, at low concentrations, which could affect the organisms exposed to it. By the *Allium cepa* test system (onion), these authors showed that the hydrocarbons found were able to induce aneugenic and clastogenicgenotoxic damages, besides inducing cell death in the meristematic cells. Besides the impacts on the diverse organisms, socio-economic impacts were also generated, since part of the local population is dependent on the eco-tourism of the region (Cetesb, 2005).

Another example of accident occurred in 2005 in the Jilin province - China, where there was the explosion of a petrochemical industry. Besides causing the death of 5 people, approximately 100 tons of a mixture of benzene, nitrobenzene and other compounds leaked in the Songhua River, which provoked the cancellation of the water supply to a city 500 km downstream the local of the accident (Fu et al., 2008).

There are similarities and differences of an oil spill in an oceanic environment and in an inland environment (lakes, ponds, rivers, streams) (Sergy; Owens, 2011). According to these authors, it must be taken into consideration the particular physical and biological characteristics of these environments, which will determine how a response will be madeto remedy the possible problems generated.

The sensitivity and productivity of geological extracts where the oil is accumulated is lower in freshwater environments, when compared to marine, which enables the use of more aggressive techniques to its removal. However, a greater attention is needed to the beds of shallow rivers, mainly in their mouth regions, near marine environments, since in these sites there are a higher activity and biological diversity. Also in relation to environments of inland waters, it is important to take into account seasonal variations that the water bodies suffers, which interferes in their flow, volume and capacity to reach floodplains, in the case of large floods, since the strategies of oil removal depend on these factors. In the marine environment there is a greater dilution and dispersion of the oil, helping in its removal and biodegradation, which does not occur in small freshwater bodies. Only in rivers with great energy (great turbulence or strong winds that generate waves) the oil removal will be facilitated by the washing that the water promotes (Sergy; Owens, 2011).

Although accidents in inland waters are less frequent when compared to oceanic accidents several regions of the world are susceptible to suffer some kind of impact caused directly by petroleum or petroleum products. Regions where there are industries of processing, transportation and refining of petroleum are those that present the greatest risks and, generally, are found near water bodies, whether for its use inside the petrochemical plant or to the transportation of their raw materials or products. Thus, the constant monitoring is needed.

There are studies that try to create models for the prevention and remediation of the accidents involving several types of environments, but they are still scarce in relation to inland waters, since there are few studies that address this issue (Danchuk; Willson, 2010).

1.3. Groundwater

Groundwater is an important source of freshwater in several countries. The National Groundwater Association of the US (NGWA) estimates the use of these waters, as drinking source, in about 75% in Europe, 47% in the US, 32% in Asia, 29% in Latin America and 15% in Australia (NGWA, 2014).

Contamination of groundwater occurs mainly by underground fuel tanks and pipelines that suffer cracks and leak (Yehet al., 2010). In the case of petroleum stations, petrochemical industries, refineries and storage tanks, the crude petroleum and petroleum products can leak slowly for a prolonged period, making high the probability of groundwater contamination (Tyagi et al., 2011).

In 1997, an accident in Australia occurred with petroleum products, where 3,000 litres of organic solvents leaked from an underground ruptured tank. One part of the solvent emerged from the soil, but great part, besides reaching a nearby river, reached the local groundwater. The impacts of this leakage were investigated by Guerin et al. (2002), who detected petroleum hydrocarbons in the groundwater of the leakage site.

In 2004, there was petroleum leak in an oil pipeline in the region of Guaecá and São Sebastião, in São Paulo State, Brazil. This accident released a mixture of petroleum that, besides affecting the local river and beach, affected mainly the groundwater. The technical report of CETESB (2005) showed that the contamination plume reached an extension of 562 to 1355 m^2, f measured from 119 monitoring wells.

Doherty and Otitoloju (2013) described an accident that occurred in May 2008, in a gas pipeline of Nigerian National Petroleum Corporation (NNPC), located in the area of the Ijegun lakes, resulting in oil spills and consequent contamination of the environment. The authors researched the residual concentration of total hydrocarbons (THC) and the total concentration of benzene, toluene methylbenzene and xylene (BTEX) in groundwater, during the period of March 2009 to July 2010. The results show high levels of THC (757.97 mg/L for groundwater), whose levels were above the maximum value allowed by WHO - World Health Organization (0.1 mg/L). According to the authors, this significant contamination by hydrocarbons was registered at 500 m from the explosion site and still observed 25 months after the incident, which arouses great concern and make necessary the remediation of the affected area to decrease the negative impact on the impacted ecosystem.

BTEX is a mixture whose presence in the environment is worrying, mainly due to its potential to contaminate groundwater (Yeh et al., 2010). Due to its high solubility and polarity, BTEX can easily reach groundwater and cause acute toxicity and other adverse effects to human health. This compound is found in all types of petroleum products, and can appear in even 20% of the volume of gasoline fraction (López et al., 2008).

When assessing environment impacts caused by oil spills in the continent, it is essential to have knowledge of the effects on the surface, as well as it is important to consider the persistence of the oil in the soil, chemical transformations and its transport to groundwater (Duffy et al., 1980).

1.4. Soil

Soil is an important environmental matrix that gives both physical and chemical support to the ecosystems, which characterizes it as fundamental for the maintenance of life in the planet. Thus, the physical and chemical conditions of the soil define its viability for the safety of its function in the ecosystem. According to Sodré et al. (2001), soil is a very specific component of the biosphere, which can serve as geochemical deposit of pollutants. Therefore, soil pollution has been recognized as an important environmental problem, which may cause risks to the environmental health.

When treating with soil contamination it is important to consider the quantity of organic matter and clay present, due to the effect of the contaminant in the primitive population of microorganisms (potential primary degraders of hydrocarbons), as well as the access to these pollutants, which will influence their degradation (Joergensen et al., 1997). Over time, hydrophobic contaminants tend to become more difficult to be removed from soil and less bio-available to microorganisms. This occurs due to the break of these contaminants, which are sequestered by the soil matrix and end up composing the organic matter of the soil; or even, by its diffusion in the nanopores present in the environment (Northcott; Jones, 2001; Huesemann et al., 2004).

Contamination of soil by petroleum and petroleum products, besides causing serious risks to the own soil, preventing its use, can also cause contamination in other matrices of the ecosystem such as for example surface waters and groundwaters, which puts at risk the potability of the water and its use in agriculture, and, thus, can lead to damages to human health, economy and ecosystem (Wang et al., 2007; Xu et al., 2006).

The introduction of oil and petroleum products in the soil can occur due to failures during the exploration process, disruption of pipelines, discharge of effluents, irrigation with water containing petroleum residues, incomplete combustion of petroleum products, accidents in railways and roads, leakages in storage tanks, landfarming and incorrect disposal in landfills (Chaîneau et al., 2000; Tang et al., 2012).

High concentrations of hydrocarbons can destroy vegetation due to their phytotoxic properties (Chaîneau et al., 1997). Thus, the presence of petroleum in the soil alters both its physico-chemical composition and the dynamics of the microbiota present in it (Hamamura et al., 2008).

Although soil contaminations are quite frequent, they are poorly reported and do not attract the media attention as accidents in aquatic environments. According to Burgherr (2007), the number of accidents in oil pipelines doubled, while the volume of oil spilled tripled.

One of the most tragic soil accidents already registered date of 1994, occurred near the city of Usinsk, in the south of the Arctic Circle, former Soviet Union. In this accident, there was the leakage of an oil pipe for about 9 months. When the leakage was detected, ditches and dams were constructed to contain and collect the oil spilled. However, with the heavy rains that occurred in that year, the dam broke, spilling more than 100,000 tons of oil into the tundra and Pechora River. Since it was located in a region of difficult access and in periods of autumn and winter, the logistics of the equipments and availability of techniques were quite limited. After the occurrence of a new leakage, whose volume was not disclosed, 350 ha of frozen area were covered by petroleum. Part of the remediation of this oil spilled was done by incineration, but great part remained in the local (Vilcheck; Tishkov, 1997; Jernelöv, 2010).

The presence of soil, associated with the presence of fire, damaged the tundra vegetation, marking the first evidence of damages occurred in the area. With the thaw, the contaminants spread out to other rivers, lakes and plains, until reaching the ocean (Vilcheck; Tishkov, 1997; Jernelöv, 2010).

In the same region (near Usinsk), it had already occurred, in 1988, the disruption of an oil pipeline, responsible for the release of 20,000 tons of oil, followed by two similar events in 1992, reaching around 30,000 tons of oil released into the environment (Jernelöv, 2010).

In Colombia, the Caño Limón pipeline, with 780 km of extension, whose purpose was to transport oil to the Caribbean coast, was target of attacks of FARC (Revolutionary Armed Forces of Colombia) and ELN (National Liberation Army), for more than 100 times. The total volume of oil spilled by this pipeline, along the years, was estimated in about 400,000 tons (Hosmer et al., 1997; Jernelöv, 2010).

Also according to Jernelöv (2010), in the Colombian, Peruvian and Ecuadorian Amazon, the oil and gas exploration activities contaminate large portions of the forest. In the 60's and 70's decades, the North American companies Occidental Petroleum Corporation active in Peru, and Texaco (now part of Chevron) in Ecuador, were charged with several environmental abuses in the Amazon territory, among them, several cases of oil spills resulted from inadequate infrastructure and insufficient maintenance.

During the first Gulf War, in Kuwait, about 60 million barrels of oil were released in 20 million m^3 of soil, forming 320 oil lakes in the desert of the region (Al-Saleh; Obuekwe, 2005).

The soil of the Antarctic region is considered extremely susceptible to damages caused by man due to extremely slow rate of biodegradation (Campbell et al., 1998). According to Cripps and Priddle (1991), the hydrocarbons derived from petroleum compose the most abundant group of contaminants in this environment. Aislabie et al. (2001) studied the effect of petroleum contamination on the soil of an Antarctic region and, in this study, the authors observed a large number of cultivable heterotrophic bacteria and fungi. Thus, the hydrocarbons derived from spills in Antarctic soils may serve as substrate for microbial growth and, if frequent or in great volumes, these spills could affect microbial populations already existent in these soils.

2. Impacts to the Environment and Biosphere

An oil spill immediately generates concern and the media, normally, discloses it as an environmental disaster. However, according to Kingston (2002), the fate of the petroleum from a spill will depend on the type of oil and amount of oil released into the environment involved. In the ocean, for example, great part of the oil spilled forms a thin surface layer on the water, facilitating the dispersion of chemical volatile compounds, generally toxic, reducing their quantities in 30 to 40%.

Even if an oil spill occurs, it will not always cause a great impact on the local biota. Kingston (2002) reports the case of the tanker *Jessica*, which spilled, near the Galapagos islands (environment of great biological importance), diesel oil, considered a light oil, of low density, where the winds and currents carried the oil spots away from the coast and, associated to the great tropical heat, almost no oil interfered in the balance of the region, despite the major impact in the other regions that received the oil.

According to Al-Majed et al. (2012), the impact caused by oil spills cause severe impacts on the affected ecosystems, and may affect the marine life such as seabirds, mainly in the reproductive period, and also in crustaceans. Still according to these authors, the control of an oil spill depends on factors such as sea conditions (wave size), speed and direction of the wind, currents and tides, temperature and atmospheric conditions. Swedmark et al. (1973) points out that different species present distinct responses, according to their behaviour, physiology and morphology. However, a same species may also react differently, when exposed to hydrocarbons, depending on the exposure period, sex, age and contamination history (NOAA, 1992).

The effects of hydrocarbons pollution depend on the species impacted. Generally, the less sensitive species are gastropods and polychaetes, while corals, bivalves, decapod crustaceans and echinoderms are the most sensitive (NRC, 1985). Therefore, oil spills accidents can result in population increase of some tolerant species or, in other cases, reduction or extinction of more sensitive species (Spies; Desmarais, 1983; Peterson et al., 1996).

Thus, impacts caused by oil contaminations can cause impacts in marine ecosystems, environments that, according to some authors (NIkitik; Robinson, 2003; Gómez Gesteira; Dauvin, 2000) can suffer impacts in their sensitive faunal groups, being susceptible to reduction or extinction of species, such as for example, amphipods. On the other hand, the tolerant or opportunistic fauna, represented by polychaetes, can present prevalence in the environment due to their population increase (Seo et al., 2014).

According to Kingston (2002), the impacts caused by oil spills in seabirds are the most likely to attract public attention. For Peterson et al. (2003), seabirds represent the group of animals that most suffer impacts after accident with petroleum, corroborating the study of Irons et al. (2000), who report that this group is intensely impacted due to their contact with the sea surface and for breeding in the coast, place where much oil accumulates.

The effects of oil on mammals and sea turtles occur in short term and can cause acute and chronic toxic effects that include skin coating (causing hypothermia, asphyxia and drowning), ingestion of toxic substances during the dive, ingestion of contaminated prey and disruption of social functions in species socially organized (Peterson et al., 2003).

Batten et al. (1998) investigated the effects of oil spills, after the shipwreck of the Norwegian tanker *Sea Empress*, in the plankton communities of the southern Irish Sea. Planktons are considered important components of the marine ecosystem. The authors suggested that the oil spill did not cause significant effects on the community studied and that some events observed are not related with the spill. However, it was detected a small change in the species composition after the accident. SEEEC (1996) reported the effects of this same accident in the Pembrokeshire coast (United Kingdom), in which the rocky and sandy shores were taken by petroleum and, as consequence, several littoral species died or were trapped along the coast. This contamination also caused a major impact on seabirds, being documented among the 6900 oiled birds of 28 different species, the death of part of them and the recovery of another part, after 3 months of the accident. Lyons et al. (1997) performed the measurement of DNA adducts to assess the genotoxic potential of the same accident on a native species of fish. These authors obtained significant results, relating the increase of DNA adducts with the possible genetic damages observed in this species.

Zenetos et al. (2004) studied the effects of oil spill after the accident with the tanker *Eurobulker* (South Evoikos Gulf), which spilled 700 tons of crude petroleum in this place. The benthic fauna community showed an immediate decline in the local biodiversity and abundance of some species, however, the environment was totally recovered after 8 months.

Le Hir and Hily (2002) monitored during one year the macrofauna and flora of three habitats of the rocky shore of a region impacted by the oil spill *Erika*. Each habitat presented specific responses to the impacts, such as: 1) changes in the abundance, but not in the biodiversity; 2) variation of the biodiversity, causing disappearance of some species and immigration of opportunistic species; 3) loss of several species and habitat destructuring.

Kingston et al. (1995) observed that there were no significant alterations in the structure of the benthic communities, such as variance in the abundance and diversity of species after the accident with the tanker *Braer* (Shetland Islands, UK - 1993), probably due to the low toxicity of the oil spilled or to the slow responses of the benthic fauna. According to Kingston (1995), the concentrations of oil in fishes of this region reduced to acceptable levels within 2 months after the spillage. The authors observed that the clams were the most affected organisms, requiring about a year to recover to acceptable contaminant levels, while the crayfishes remained contaminated, probably because they live in the sediment. The impact on the habitat was minimal and most part of the coast was recovered within a year.

Gomez Gesteira et al. (2003) compared the effects of oil spills on the macrofauna abundance, after the oil spill in the Aegean Sea (Galicia, Spain) and of *Amoco Cadiz* (Brittany, France). The authors observed that both regions presented significative absence of opportunistic species proliferation and also the disappearance of amphipods, with low recolonization rates. Samain et al. (1980) observed a high mortality of the zooplankton in the affected areas, few weeks after the *Amoco Cadiz* spill.

Peterson (2001) studied the effects of the *Exxon Valdez* oil spill (Alaska) on several ecosystems. After the spillage, it was observed effects on several habitats and on various species. It was observed that some species died or were removed from this habitat, such as for example, snails and mussels, while there was recolonization of the shore by ephemeral algae and opportunistic species. A decrease in the abundance of several amphipods sensitive to toxic substances present in petroleum was observed, which was not recovered after the normalization of the environmental conditions. Birds and fishes were also affected, were there was a noticeable reduction in the number of species, although these groups have undergone a recovery process.

After six months of the *Prestige* shipwreck, Junoy et al. (2005) studied the effects of oil spill on 18 beaches, located along the 1659 km of the Galicia shore in order to investigate the impacts that the fuel and the cleaning procedures would cause in the endemic macrofauna. All the beaches studied were contaminated by the fuel and were considered polluted and some of them were completely covered by petroleum. A decrease in the abundance of the macrofaunas was observed, while rare species were eliminated from the environment and the macrofauna was numerically dominated by amphipod crustaceans.

According to Barros et al. (2014), seabirds are one of the most affected organisms when there is an oil spill. The authors observed that the species *Phalacrocoraxaristotelis* (European Shag) suffered great population impact at the population level after the Prestige accident and also presented reproductive deficiency.

Seo et al. (2014) evaluated, during 3 years, the initial impacts of the *Hebei Spirit* oil spill in the macrobenthic communities of the shallow coastal ecosystems that were impacted by this accident. The most affected areas presented reduction in the abundance of the sensitive benthic fauna and increase of the opportunistic fauna. On the other hand, some beaches did not suffer severe impacts and showed that the biodiversity and density of the species were recovered a few months after the spillage. Jung et al. (2011) observed that the PAH concentration in the fish tissues decreased to acceptable levels, although the concentration of PAH metabolites in bile and the induction of CYP1A in the liver took a longer time to return to normality. Moreover, studies showed that the genotoxicity and the endocrine effects caused by the contaminants present in the sediments of the *Hebei Spirit* spill area remained for over two years ([illegible] et al., 2011; Hong et al., 2012).

Incardona et al. (2014) assessed the effects of the petroleum released by *Deepwater Horizon*, in the Gulf of Mexico, in commercially and ecologically important species of pelagic fishes. Three distinct species of pelagic fishes, which presented cardiotoxicity in the form of heart failure and oedema indicating sensitivity to petroleum, were studied.

According to Kostka et al. (2011), the explosion of the *Deepwater Horizon* platform caused great impact on the composition and abundance of the native bacteria community of the Gulf of Mexico. Sand samples of the Pesacola beach, in Florida, which was severely hit by the oil from the spill, were collected. The authors identified and characterized the degradation potential by native bacteria and showed the presence of the bacterial classes Gammaproteobacteria (Alcanivorax, Marinobacter) and Alphaproteobacteria (Rhodobacteraceae) as the main responsible for the degradation of the oil present in the region.

3. Occupational Exposure

Few studies have addressed the risks that oil spills and cleanup activities can bring to human health. The *Exxon Valdez* spill was one of the first accidents that investigated the health of the people involved. The psychological effects, such as anxiety and depressive syndromes, were the most described symptoms in the area affected by this accident (Palinkaset al., 1992; 1993).

A study performed by Jung et al. (2013) investigated the respiratory effects in 436 children living along the Yellow Sea, South Korea, after the *Hebei Spirit* spill. All the evaluated were subjected to a battery of tests and were divided into two groups: the ones that lived near and the ones that lived far from the spill area. This study showed that the children living near the affected area presented more evident symptoms of asthma, increased lung function and increased prevalence of bronchial hyperresponsiveness, when compared with the children living in more remote areas.

Sim et al. (2010) also evaluated the health status of the people involved in cleaning the *Hebei Spirit* oil spill and identified the risk factors associated with the development of the symptoms. The symptoms were classified into six categories: backache, skin lesions, headache and pain in the eyes, neurovestibular alterations and respiratory effects. The authors concluded that this accident was related with the health problems found in the residents and volunteers that participated in the cleanliness of the place and the symptoms were directly proportional to the exposure of the individuals.

Suaréz et al. (2005) studied the acute effects in the health of individuals who were involved in the cleaning of the black tide generated by the accident with the tanker *Prestige* (northwest of Spain). Four different groups participated in the study: bird cleaners, volunteers, workers and sailors. No serious disturbances were observed among the individuals that participated of this research; however, the group of sailors was the most affected. The authors concluded that the acute effects on the health are related to the type of activity developed and observed that the performance of several activities increases the risk of lesions. Now, Rodríguez-Trigo et al. (2010) assessed the respiratory effects and the chromosome damages in volunteers, including fishermen, 2 years after these people participated of the cleanup activities of the accident with the tanker *Prestige*. The individuals that participated of the cleaning of this region presented persistent airway lesions and, moreover, presented structure chromosome aberrations in circulating lymphocytes. Laffon et al. (2006) carried out a study to evaluate the genotoxic effects (comet assay, micronucleus test and sister chromatid exchange test) in individuals exposed to petroleum, whose results showed a significant increase in the DNA damages. After 7 years, Laffon et al. (2014) developed a study to verify the possible persistence of genotoxic damages in individuals exposed to the petroleum of *Prestige*. The obtained results showed that there was no evidence of DNA damages among the individuals studied.

Crum (1993) documented that children that lived near the oil spill area caused by the shipwreck of *Braer* did not present impairment in their lung functions, immediately after the spill. Campbell et al. (1993) reported that the most frequent symptoms found in the individuals who came into contact with the oil spilled by the tanker *Braer* were eye irritation, headaches and sore throat. Cole et al. (1997) evaluated that the damages to the DNA of the population exposed, by genotoxicity tests, and observed that the oil spilled did not present genotoxic potential.

Meo et al. (2008) evaluated the lung function in individuals exposed to oil spill after the accident with the tanker *Tasman Spirit,* in Pakistan. It was verified that the individuals exposed to petroleum presented impaired pulmonary function. However, this deficiency was reversible after one year of the accident.

Morita et al. (1999) analyzed the acute effects in 282 people that helped in the physical removal of petroleum type C, spilled in the coast of Japan by the tanker *Nakhodka*. After the performance of the urianalysis of the involved and the analysis of the concentrations of petroleum products in the environment, the authors concluded that some of the reported symptoms, such as headaches, eye and mucosal irritation, may be associated with exposure to oil. However, it was not registered any significant concentration of petroleum products in the air that could induce such effects in the exposed individuals.

Lyons et al. (1999) investigated if the dwellers of the region affected by the *Sea Empress* oil spill presented an increase in the physical and psychological symptoms that could be due to exposure to the crude oil. The authors observed that the high rates of anxiety, depression, headaches, pain in the eye and sore throat of the individuals analyzed were associated to areas with higher oil load.

Due to the great number of accidents occurred throughout the history of the petroleum industry, the great influence of this compound on the local biota and on human beings is very important to the secure remediation of the environment.

4. Remediation Methods

Currently, to try to remove the oil spilled, there are several methods that can be separated into physical (among them the thermal ones), chemical and biological techniques. We can also divide the remediation methods in *in situ* and *ex situ*, where the *in situ* remediation is performed on the site where there is the contaminated material, thus, considered more effective and economically viable and the *ex situ* remediation that is based on the removal of the contaminated material to be treated in another place and, therefore, considered more expensive and difficult to be carried out, having, even, a possibility to spread the contamination (Farhadian et al., 2008). The most used techniques to remove petroleum, such as the oil *in situ* burning in the water, booms, skimmers and even natural inorganic and synthetic adsorbents, are often ineffective at high tides, expensive and are harmful to the environment (Al-Majedet al., 2012).

The choice of the most efficient method will depend on the type and amount of oil spilled, besides the conditions of the environment in which there was the spillage. Thus, according to Dave and Ghaly (2011), it is necessary to understand the amount and the chemical and physical properties of the petroleum, its age, climate characteristics and the environment involved, so that the best remediation method to be used be correctly chosen.

Before the remediation methods are employed, the petroleum may undergo some alterations that can interfere in its degradability. According to Al-Majed et al. (2012), when an oil spill occurs on the sea surface, this petroleum may suffer, simultaneously, several processes such as spreading, evaporation, emulsification, photooxidation, dispersion, immersion, resurgence, formation of tar balls and biodegradation, where they can all make the cleanup of this oil even more difficult.

The UV radiation from sunlight is able to oxidize some compounds present in the petroleum by a process called photolysis. The products derived from this oxidation include acid and phenolic compounds and, although many of them are more toxic than the original hydrocarbons, the concentration of these compounds is very low to be considered ecologically significant (Kingston, 2002).

Some hydrocarbons, generally the ones with low molecular weight that are relatively toxic, dissolve into the seawater (dissolution) and this process is responsible for a small percentage (<1%) of the degradation of the hydrocarbons of the petroleum spilled (Kingston, 2002). Still according to this author, the process called dispersion is probably the responsible for the natural removal of most part of the petroleum spilled, where the oil is broken by the action of the sea waves into small drops (0.01 to 1.0 mm in diameter), which are retained in the water column until being degraded by the action of bacteria.

Besides the natural remediation processes and dispersion of the oil spot, more effective methods were developed.

4.1. Physical Methods

According to Dave and Ghaly (2011), the physical methods are commonly used in oil spills that occur in aquatic environments. These methods function as containment barriers to avoid that the petroleum spreads without changing its physical and chemical characteristics. Among these barriers we can cite: booms, skimmers and adsorbent materials.

Booms are floating barriers that contain the oil spreading and can be divided into three categories: fence boom, curtain boom and fire resistant boom. Skimmers are machines used in the remediation of oil spills that can separate liquid particles in suspension as well as separate one liquid from another.

The adsorbent material can be applied together with skimmers in the petroleum removal process, since they can transform the liquid phase in semi-solid phase (Teas et al., 2001) and, therefore, facilitate the oil removal. The origin of these hydrophobic adsorbents can be natural (organic or inorganic) or synthetic.

Still according to Teas et al. (2001), the synthetic adsorbents are the most marketed and used and can be produced from polypropylene, polyester and polystyrene foam, in the form of sheets, rolls and dams, or even be used on the water surface in the form of powder. Dave and Ghaly (2011) affirm that the adsorption capacity of these materials is about 70 to 100 times their weight due to their hydrophobicity and lipophilicity. However, according to the same authors, the difficult to store and degrade these compounds are disadvantages of this method. Added to these disadvantages, there is the fact that these adsorbents are not advisable for use on water surfaces, since they are difficult to be applied in windy days and can present risk if inhaled.

Among the natural inorganic adsorbents we can include clay, glass, sand, vermiculite and volcanic ashes (Holakoo, 2001). Amid the natural organic adsorbents, straw and cotton are the most promising, although they cannot be available in certain regions. Thus, the use of alternative materials such as coconut shells, corn cobs, sugarcane bagasse, human and animal hair have been encouraged as well as other forms of natural adsorbents (Al-Majed et al., 2012), since they can have high absorption capacity.

4.2. Thermal Method

This method is based on the *in situ* burning of the petroleum, and is considered a simple and highly efficient technique to eliminate large amounts of oil, and it is also considered fast and inexpensive, however, this technique is limited by some factors such as wind and sea conditions and presents as a disadvantage the formation of a black smoke that is released into the atmosphere (Allen; Ferek, 1993).

4.3. Chemical Methods

The chemical methods are based on the use of substances belonging to the chemical classes of dispersants and solidifiers.

According to Lessard and Demarco (2000), dispersants are surfactants (agents that act on the surface tension) and act similarly to detergents, these substances are used on oil spots to remove them from the water surface and disperse them in the water column, decreasing their concentrations and accelerating the oil degradation by natural processes. Still according to Lessard and Demarco (2000), surfactants are designed to have affinity for both the oil (lipophilic portion) and water (hydrophlic portion) and, when applied on the oil film, diffuse in the oil/water interface; posteriorly, they realign and the lipophilic portion binds to the oil phase and the hydrophilic portion extends to the aqueous phase, therefore, this process reduces the interfacial surface tension between water and oil, making that the oil remains in the form of small droplets.

According to Al-Majed et al. (2012), the effect of chemical dispersants normally used to control oil spills can be even more harmful than the effect caused by the oil spill, since it can take the oil to places where it would not go naturally and facilitate its entrance into several organisms.

Solidifiers are composed by dry granular material (hydrophobic polymers) that react with petroleum forming a solid and cohesive mass that floats in the sea (Adebajo et al., 2003), when the petroleum solidifies it can be easily removed leaving very little residue (Ko; Day, 2004).

4.4. Biological Methods

Biodegradation consists of a process in which microorganisms (bacteria, yeasts and fungi) naturally transform complex compounds in smaller compounds during their feeding process (Al-Majed et al., 2012). Still according to these authors, the biodegradation process can be used in the removal of oil spilled, by the biostimulation technique, which is based on increasing the native microbial activity by adding nutrients in the environment; by the bioaugmentation technique, which is based on the addition of exogenous degrading microbial strains; and, by the combination of both.

For Atlas (1991), an efficient bioremediation must convert hydrocarbons into non toxic substances, such as water and carbon dioxide. When the metabolic processes are used to remediate pollutants, without any alteration in the natural conditions of the environment, they can be called natural attenuation or intrinsic attenuation (Farhadian et al., 2008).

Although the bioremediation is a technique that allows the application of several methodologies, there is a common fundamental principle between all of them: the use of microorganisms as basis of the hydrocarbons degradation process (Daane et al., 2001). These hydrocarbons are excellent carbon sources and, consequently, energetic to the degrading organisms, such as prokaryotes and eukaryotes (Prince et al., 2010; Prince, 2010). Despite these organisms are able to initiate the biodegradation process, in the oil spilled there is lack of viable sources of nitrogen, phosphorous and other essential elements to the microbiological development (Prince; Butler, 2013). Fortunately, when hydrocarbons are found at low concentrations, the own salty water of the seas is able to provide microorganisms the necessary amounts of these nutrients (Prince et al., 2013).

The bioremediation process has some limitations, since it is not efficient to remove large amounts of oil from the environment or remove oils derived from spills that sank in the sea (Wardley-Smith, 1983). Moreover, the bioremediation process is limited by abiotic environmental factors, such as low levels of nutrients (e.g., phosphates and fixed forms of nitrogen), very low temperatures and insufficient quantities of oxygen (Atlas; Cerniglia, 1995). According to Atlas (1991), the physical state of petroleum hydrocarbons also influences in their biodegradation.

Temperature interferes in the biodegradation rate by affecting the physical and chemical properties of the petroleum and at low temperatures the petroleum viscosity increases, volatilization of toxic short-chain alkanes decreases and its solubility in water enhances, making that this biodegradation process be retarded (Atlas, 1991).

Aromatic substances of high molecular weight, resins and asphaltenes are considered recalcitrant or exhibit very low degradations rates (Atlas, 1991). Biodegradation of aliphatic hydrocarbons or light aromatic fractions occur more easily than aromatics, aliphatics of normal chain are more quickly than the ones of branched chain, saturate more than unsaturated and the ones with chains between 10 to 20 carbons are also more easily degraded (Pereira Jr., et al., 2009).

According to Van Hamme et al. (2003), the biodegradability of the petroleum compounds occurs more easily from: n-alkanes > branched chain alkanes > branched alkenes> n-alkyl aromatics of low molecular weight >monoaromatics> cyclic alkanes > polycyclic aromatic hydrocarbons (HPA) >asphaltenes.

As previously mentioned, there are *in situ* and *ex situ* techniques for remediation. As the most used *in situ* remediation techniques we can cite, according to Vidali (2001):

- bioventilation: process in which there is the injection of air currents on the soil subsurface (aiming to treat contaminants dissolved in the groundwater or sequestered within the soil pores) and addition of nutrients through wells in the contaminated site (to stimulate the autochthonous bacteria). However, this method only serves for the remediation of simple hydrocarbons;
- *in situ* biodegradation: process in which there is the supply of oxygen and nutrients by aqueous solutions that circulate in the contaminated site, in order to stimulate the bacteria that occur naturally in the site to degrade the organic contaminants. This technique can be used both in soils and groundwaters;
- biosparging: consists in injecting air under pressure below the water table to increase the concentration of oxygen existent in the groundwater and, thus, increase the biodegradation rate by autochthonous bacteria. Biosparging increases the mixture in the saturation zone and hence enhances the contact between soil and groundwater;
- bioaugmentation: consists in the addition of native or exogenous microorganisms in the contaminated sites.

Still according to Vidali (2001), the most commonly used *ex situ* bioremediation techniques are:

- *landfarming*: is a simple technique, where the contaminated site is excavated and spread on a prepared "bed", which is periodically turned over until the pollutants degradation. The objective of this technique is to stimulate the native biodegrading microorganisms and facilitate the aerobic degradation of the contaminants;
- composting: in this technique there is the combination of contaminated soil and non-hazardous organic correctives (e.g., manure and agricultural residues). The presence of these substances helps in the development of a microbial population and increases the temperature of this mixture;
- biopiles: it is a hybrid between landfarming and composting, where aerated cells are constructed, similar to the composting piles, in order to stimulate the aerobic microbial activity within the pile. The microbial activity can be increased by adding moisture and nutrients such as nitrogen and phosphorous;

- bioreactors: mud or aqueous reactors are used for the ex situ treatment of contaminated soil and water pumped from the contaminated plume. This technique requires the processing of the contaminated material by a contention system, which creates conditions to increase the bioremediation rate.

The most conventional methods used to remove gasoline compounds from contaminated groundwater include aeration (air stripping), absorption with activated carbon, filtration and biodegradation (Garoma et al., 2008). Besides these, there are more developed methods such as oxidation made with ozone, hydrogen peroxide, ultraviolet rays and semiconductors (for example, titanium oxide - TiO_2) (Mascolo et al., 2008).

According to Andrade et al. (2010), several microorganisms can be used in the remediation process (bacteria, filamentous fungi and yeasts) and the most used are bacteria. Bacteria with potential to degrade hydrocarbons are naturally present in several ecosystems, from polar soil until marine environment (Daane et al., 2001).

Pacwa-Plociniczak et al. (2014) assessed the efficiency of the *Pseudomonas* sp. P-1 strain in degrading hydrocarbons and produce biosurfactants. The authors verified that this strain was able to degrade crude oil, A5 and P3 fraction of the crude oil and hexadecane, besides being efficient in producing rhamnolipids (biosurfactants) and this characteristic is very important, since these compounds increase the bioavailability of the hydrocarbons. Thus, the authors concluded that the combination of ability to produce biosurfactants with the ability to degrade hydrocarbons makes that the use of *Pseudomonas* be one of the most useful tools to bioremediate soils contaminated with petroleum.

Qin et al. (2013) evaluated the effect of biostimulation corrected with biochar (carbonized biomass) in the bioremediation of soils contaminated with petroleum. In this study the authors verified that after biostimulation there was an increase in the abundance of several groups of bacteria, but with the bioremediation process there was a decrease in this abundance. In an attempt to correct this situation, the authors used the biochar derived from rice straw and verified that this addition increased the efficiency of the contaminant degradation, when compared with soils only biostimulated. Thus, the biochar derived from rice straw can be used for the remediation of soils contaminated with petroleum and can act as an effective additive in soil remediation.

Gargouri et al. (2013) assessed the efficiency of the bioaugmentarion process in the bioremediation of a contaminated refinery soil, located in a semi-arid region of Tunisia, and analyzed the response of the native and exogenous bacterial communities in the presence of hydrocarbons. Initially, the authors analyzed the amount of native bacteria before the bioaugmentation process. To verify the bioaugmentation efficiency a bacteria consortium, obtained from a reactor used in the treatment of residual waters of the refinery, was inoculated in the soil, the obtained results were compared with a control soil, without the consortium. The results indicated a considerable increase in the number of bacteria, after the bioaugmentation, which culminated in an increased degradation of the hydrocarbons. Thus, the authors concluded that the bioaugmentation is effective to remediate contaminated soils in semi-arid regions.

Gomez and Sartaj (2013) evaluated the efficiency of the bioaugmentation and biostimulation process in biopiles in the remediation of soils of cold climates. The authors used soil with hydrocarbons from *Val-des-Bois* (Quebec, Canada), which was excavated and transported to Moose Creek (Ontario, Canada). With this soil biopiles were constructed, in which different treatments were performed: only soil, soil+microorganisms consortium (commercial product containing a mixture of native bacteria strains from Canada, with ability to degrade hydrocarbons), soil+compost (containing organic matter, phosphorus, nitrogen and carbon), soil+consortium+compost. All piles received oxygen and carbon dioxide supply. The authors observed that there was a higher degradation in the biopiles that have the microorganism consortium associated with the compost (bioaugmentation+biostimulation), with an efficiency of 82% in the removal of hydrocarbons. In the piles that have only soil with the microorganisms consortium, (bioaugmentation) or only soil with the compost (biostimulation) there was also a higher biodegradation than in the soil without any additive, but not so efficient. Thus, the combination of bioaugmentation and biostimulation showed to be an important method in the remediation of sandy soils of cold climate.

Yeh and collaborators (2010) describe that permeable reactive barriers are one of the most effective forms to remediate an environment with groundwater contaminated by BTEX and other organic compounds. Contaminated groundwater cross transversely the barriers, allowing a physical absorption of the contaminants, which can be chemically or biologically degraded. When biodegradation happens, in this case the aerobic metabolism is the most effective in the contaminants removal, however groundwater is poor in dissolved oxygen and it is necessary the use of materials and/or substances that release oxygen in the water. Still, even the aerobic biodegradation being more effective, it is not always possible the direct injection of oxygen, making the anaerobic biodegradation the only alternative (Farhadian et al., 2008).

Bioremediation, performed by the biodegradation by microorganisms for the removal of contaminants of groundwater, will depend on the type of aquifer and soil (Farhadian et al., 2008). According to these authors, this method has a relative simplicity, low cost and proven effectiveness in removing pollutants. Still according to the authors, *in situ* bioremediation is a process where the microorganisms metabolize the organic compounds and transform them into inorganic substances, like carbon dioxide, methane, water and inorganic salts.

Farhadian et al. (2008) gathered information from the literature on the factors that affect the bioremediation of the groundwater contaminants: origin and concentration of the contaminant; chemistry and toxicity of the contaminant; solubility, transport, adsorption, dispersion and volatility of the contaminant components; chemistry, physics and microbiology of the groundwater; hydrogeology and hydrology of the contaminated site; limitations of the environmental patterns for water and soil; environmental conditions, sources of nutrients and presence of electron acceptors; and, finally, biodegradability of the contaminants and presence of a microorganisms population that biodegrade.

Mazzeo et al. (2010), in *in vitro* studies of aerobic biodegradation of the BTEX, observed that this biodegradation occurred by a consortium of different microorganisms, exclusively of five bacteria species, being the bacteria *Pseudomonas putida* the responsible, in this case, for the BTEX biodegradation. This was confirmed by chemical degradation analysis, which showed that BTEX were reduced in biodegraded concentrations.

A study carried out by Chen et al. (2010) evaluated the efficiency of the natural attenuation and biostimulation in the control of bioremediation of groundwater contaminated with BTEX. Although natural attenuation is an important process to contain and decrease the contamination plume, biostimulation, by air injection, showed to be a more effective and faster process. The air laden with oxygen changed the anaerobic degradation profile to aerobic, making the removal of BTEX increase to 99%. The authors concluded that natural attenuation and the stimulated aerobic bioremediation are important and effective tools for the remediation of groundwater contaminated with BTEX.

Winquist et al. (2014) evaluated the effect of fungi bioremediation in soils contaminated with PAH and determined that the species *Phanerochaete velutina* was the most effective in the bioremediation process. In the tests performed in laboratory, the soils contaminated with PAH, inoculated with *P. velutina*, had, in only 3 months of biodegradation, a reduction of 96% in the total quantity of PAH. The authors also observed PAH degradation in the control soils without the fungal inoculums by native microorganism but when the inoculums were added the degradation was higher. In the field experiments, the contaminated soil was mixed with composted green waste, in the form of biopiles, in the presence and absence of the inoculums. It was verified that the amount of native bacteria was the same in the fungal treatment pile and in the control pile, as well as the percentage of PAH degradation (approximately 94%, in the three months). According to the authors, this difference in the biodegradation between the laboratory and field tests (96 and 94%) could be the result of a lower concentration of PAH in the field experiment, reducing the toxicity and enabling the degradation by native bacteria.

Corroborating the results obtained by Winquist et al., (2014), Mohsenzadeh et al. (2012) evaluated the efficiency of four strains of fungi (*Acromonium sp.*, *Alternaria sp.*, *Aspergillus terreus* and *Penicillium sp.*) in the bioremediation of soils contaminated with petroleum. These four strains were chosen for this study because they are frequent in areas contaminated by petroleum. The authors observed that all four strains were able to grow considerably in soils contaminated with a small amount of petroleum and were able to grow much less in very contaminated soils. These results show that these strains can be used in the remediation of contaminated soils, preferably with low concentrations of petroleum. Fungal enzymes were essential in the contaminants degradation and fundamental in the soil bioremediation (Mohsenzadeh et al., 2012).

Tang et al. (2012) evaluated the efficiency of the degradation of crude oil by an artificial consortium between one microalgae and a pool of 4 bacteria. The artificial consortium was done with an axenic alga *Scenedesmusobliquus* (which does not present capacity to degrade petroleum, but promotes significantly the degradation ability of bacteria) and four degrading bacteria (*Sphingonomas, Bulkholderiacepacia, Pseudomonas, Pandoraeapnomenusa*). The authors observed that the consortium was able to completely eliminate alkanes in 10 days, alkylcycloalkanes and alkylbenzenes in 7 days, besides degrading other PAH such as naphthalene, fluorene and phenanthrene.

Several researches have isolated yeasts from different environments, including the terrestrial and marine environments, capable of degrading hydrocarbons (Manee et al., 1998; Zinjarde; Pant, 2002). Yeasts that degrade hydrocarbons are known to be producers of emulsifying substances able to form micelles or oil microdroplets (Hassanshahian et al., 2012). Moreover, several studies have investigated and related the hydrophobicity of the cell surface and biodegradation of crude oil by microorganisms (Ljah, 1998).

Hassanshahian et al. (2012) identified some yeast strains that were isolated from environments contaminated by petroleum in the Persian Gulf and assessed their capacity to biodegrade the components of crude oil, besides evaluating the emulsification activity. Among the isolated, two strains, identified as *Y. lipolytica*, were able to degrade aliphatic hydrocarbons efficiently. However, they were not capable of growing in the presence of aromatic hydrocarbons. Furthermore, the authors report a direct relationship between the cell hydrophobicity of the strains and their emulsification activity.

Abioye et al. (2013) isolated the species *Saccharomyces cerevisiae* from a brew and tested, for 28 days, the efficiency os this yeast to degrade, under aerobic conditions, the crude oil. At the end of the experiment, the authors observed that 49.29% of the crude oil was degraded. Thus, the results suggest that the use of this yeast can be effective to recover environments contaminated by petroleum.

Mattew et al. (2008) isolated and characterized strains of hydrocarbon-degrading yeasts present in a pond water of Nigeria. The authors identified two strains of Candida (*C. albicans and C. maltose*) as yeasts able to degrade petroleum.

Despite the bioremediation is proven effective when performed with microorganisms, there are other organisms that can collaborate with the remediation of degraded areas. Phytoremediation is an emerging green technology that uses plants to remediate soil, surface water and groundwater contaminated with substances such as metals and organic compounds (Pradhan et al., 1998).

Due to their sessile nature, plants have the ability to withstand to a wide range of alterations of the environment (Wang et al., 2008). According to Ndimele (2010), several plant species have shown ability to grow in contaminated environments and extract from them pollutants. Phytoremediation is carried out efficiently and can become an alternative form to remediate impacted environments, since it is a noninvasive treatment that can be occur *in situ* and is cheaper than the other remediation methods. However, as such approach is new and is not fully understood, it still has some limitations in the identification of effective species, and from the several species studied, grasses and leguminous are the most indicated for the petroleum removal.

According to James and Strand (2009), in the phytoremediation we can have *in planta* and *ex planta* processes. The *in planta* processes include situations in which the compounds of low molecular weight can be: absorbed from the soil and eliminated by the leaves through evapotranspiration process; degraded by enzymes (phytodegradation) or stored in non-bioavailable forms (phytoextraction), in this case the compounds are not volatile; remain stable inside the plant; or be removed together with the plant for posterior treatments, such as incineration (Gerhardt et al., 2009). Now, the *ex situ* degradation, can happen via rhizosphere (microbial activity at this location) or excretion of proteins or co-factors (James; Strand, 2009). Still according to these authors, the *ex planta* processes include the use of genetically modified plants to over-express genes of extracellular enzymes that can degrade the contaminants.

Nedunuri et al. (2000) studied the phytoremediation efficiency of 3 plant species (*Sorghum biocolor* L., sorghum; *Lolium annual*, ryegrass and *Stenotaphrum secundatum*, St. Augustine grass) in the degradation of total petroleum hydrocarbons (TPH), in a southeast area of Texas contaminated with these substances. In this study, they verified that the reduction in the mean concentrations of TPH in the areas that contained ryegrass (*L. annual*) and St. Augustine grass (*S. secundatum*) were 25% higher than in the areas that had sorghum and in the control region, which did not have plants, showing that ryegrass and St. Augustine grass were effective in the phytoremediation of TPH.

Mohsenzadeh et al. (2010) conducted a field study in a site contaminated with petroleum to find species of plants that were resistant to this pollutant, as well as strains of fungi associated with their roots. The authors found seven species of plants and 11 species of fungi and, from these plant species they selected the species *Polygonum avicular* to perform the phytoremediation. After six months of study, the authors concluded that the combined use of *P. avicular* with fungi strains associated to its roots, mainly the species of the genus *Fusarium*, was more effective to remove petroleum from the soils. Moreover, the results of this study showed that the fungi strains played a major role in the bioremediation of soils contaminated with petroleum, but the roots increased this process.

It is noted that an effective remediation of the environments contaminated by petroleum will depend both on the characteristics of the environment and of the petroleum and that, although there are several methodologies to remove this contaminant, none of these methods is totally efficient, and several methods should be tested together to promote a better remediation and cause less impact on the already impacted environments.

CONCLUSION

Fossil fuels are the major generators of energy in the world. Among the advantages of the use of these fuels, is the fact of having a facilitated extraction both in the continents and in the seas, has economically viable storage and transport and are used as raw material for thousands of products of contemporary use. Thus, we have that the man of today is dependent on this energy resource, whether it is for his maintenance or to give support to the development of new technologies that will sustain the advances and progresses in the several areas of operation and knowledge.

Among the existing options for supplying energy originated from fossil fuels there is the petroleum, considered as a fundamental fuel for the essential activities today. Thus, the use of petroleum as an energy source is seen nowadays as advantageous due to several factors such as the ease of obtaining and efficiency of achieving energy, due to the fact that it is very energetic. However, the use of petroleum also presents several disadvantages, mainly for being a compound with great pollution potential. Therefore, there is a major concern to assess the potential risks of petroleum on the environment, as well as develop new methodologies that can mitigate the possible environmental damages that this mineral resource may cause to nature and, consequently, to environmental health.

Since it is a product with restrict presence in few regions, has finite volume and generates high pollution levels, man search for alternatives to substitute the use of this mineral resource, although it is not possible or viable yet to perform its complete substitution.

Since the beginning of petroleum use by humans, it was verified its capacity to alter the environment, which allowed an accumulation of knowledge on this subject. However, many times the experiences acquired are based only on events experienced in major accidents. Images of birds and other animals covered by oil gain great prominence in the news, and are used as symbol of the contamination and environmental impact. However, we see that many contamination cases occur in a more silent way, being, therefore, less exploited or documented by media, although they can reflect on equally or more severe environmental impacts than those that receive attention and prominence.

From all the above, there is, today, a great concern about the several contaminations that can occur with petroleum, which is stimulating the development, in parallel, of both extraction and remediation techniques of this compound in the environment. New emergency strategies to contain leakages, as well as new remediation techniques have been implemented in the last decades in an attempt to minimize the impacts caused by oil spills in the environment. Currently, we see that the use of living organisms in the remediation process, mainly microorganisms, is one of the best strategies for reversing the large environmental impacts caused by chemical agents. Thus, due to this new trend of removing contaminants derived from petroleum, we see that bioremediation is, undoubtedly, the most appropriate form to reverse an impact caused by this chemical agent, possible to be applied even in different ecosystems.

The advantages of using this biotechnology, whether it is by biostimulation or by bioaugmentation, are still reinforced by its low costs and high efficiency in removing these contaminants.

However, there is still the need to develop studies aiming a better understanding of the behaviour of the microorganisms in the degradation of petroleum and petroleum products. These studies may warn about the efficiency of using this methodology or the dangers that it can represent, ensuring, or not, the efficiency of its application in the prevention of environmental damages.

There is still a lot to be studied and developed in the field of petroleum industry, mainly on the aspect of environmental preservation, so that no human influence can jeopardize direct or indirectly the ecosystems and especially environmental health.

References

Abioye, O. P.; Akinsola, R. O.;Aransiola, S. A.; Damisa, D.,andAuta, S. H. (2013). Biodegradation of Crude Oil by *Saccharomyces cerevisiae* Isolated from Fermented Zobo (Locally Fermented Beverage in Nigeria).*Pakistan Journal of Biological Sciences*,16, 2058-61.

Aislabie, J.; Fraser, R.; Duncan, S., and Farrell, R.L. (2001). Effects of oil spill on microbial heterotrophs in Antarctic soils. *Polar Biology*, 24, 308-313.

Allen, A. A. and Ferek, R. J. (1993) Advantages and disadvantages of burning spilled oil. *International Oil Spill Conference Proceedings*, 1993, 765-772.

Adebajo, M. O.; Frost, R. L.; Kloprogge, J. T.; Carmody, O., and Kokot, S. (2003). Porous materials for Oil Spill Cleanup: A Review of Synthesis and Absorbing Properites. *Journal of Porous Materials*, 10, 159-170.

Al-Majed, A. A.; Adebayo, A. R. and Hossain, M. E. (2012). A sustainable approach to controlling oil spills. *Journal of Environmental Management*, 113, 213-227.

Al-Saleh, E. S. and Obuekwe, C. (2005) Inhibition of hydrocarbon bioremediation by lead in a crude oil-contaminated soil. *International biodeteriorationand biodegradation*, 56, 1-7.

Andrade, J. A.; Augusto, F. andJardim, I. C. S. F. (2010).Biorremediação de solos contaminados por petróleo e seus derivados. *Eclética Químca*, 35, 3-17.

Annunciado, T. R.; Sydenstricker, T. H. D. andAmico, S. C. (2005). Experimental investigation of various vegetable fibers as sorbent materials for oil spills. *Marine Pollution Bulletin*, 50, 1340–1346.

Atlas, R. M. and Cerniglia, C. E. (1995). Bioremediation and petroleum pollutants. *BioScience*, 45, 332-339.

Atlas, R. M. and Hazen, T. C. (2011). Oil biodegradation and bioremediation: a tale of the two worst spills in US history. *Environmental science and technology*, 45, 6709-6715.

Atlas, R. M. (1991) Microbial hydrocarbon degradation—bioremediation of oil spills. *Journal of Chemical Technology and Biotechnology*, 52, 149–156.

Atlas, R. M.; Boehm, P. D. and Calder, J. A. (1981).Chemical and biological weathering of oil, from the Amoco Cadiz spillage, within the littoral zone. *Estuarine, coastal and shelf science*, 12, 589-608.

Barros, A.; Alvarez, D.andVelando, A. (2014). Deficiência reprodutiva de longo prazo em uma ave marinha após o Prestige derramamento de óleo. *Biology Letters,* 10, 20131041.

Batten, S. D.; Allen, R. J. S. and Wotton, C. O. M. (1998). The Effects of the Sea Empress Oil Spill on the Plankton of the Southern Irish Sea. *Marine Pollution Bulletin,* 36, 764-774.

Burgherr, P. (2007). In-depth analysis of accidental oil spills from tankers in the context of global spill trends from all sources. *Journal of Hazardous Materials,* 140, 245-256.

Campbell, D.; Cox, D.; Crum, J.; Foster, K.; Christie, P., and Brewster, K. (1993).Initial effects of the grounding of the tanker Braer on health in Shetland. *The Shetland Health Study Group BMJ*, 307, 1251–1255.

Campbell, I. B.; Claridge, G. G. C.; Campbell, D. I., and Balks, M. R. (1998).The soil environment of the McMurdo Dry Valleys, Antarctica. In: Priscu, J.C. (ed.) *Ecosystem dynamics in a polar desert: The McMurdo Dry Valleys, Antarctica*. American Geophysical Union, Washington, DC,297-322.

Cardoso, A. M. (2007). *Sistema de informações para planejamento e resposta a incidentes de poluição marítima por derramamento de petróleo e derivados*. Universidade Federal Do Rio De Janeiro.

CETESB. (2005). *Relatório Técnico da CETESB - Acidente com o oleoduto OSBAT - Petrobras/TRANSPETRO*. p. 1–101.

Chaineau, C. H.; Morel, J. L. and Oudot, J. (1997). Phytotoxicity and plant uptake of fuel oil hydrocarbons. *Journal of Environmental Quality*, 26, 1478-1483.

Chaîneau, C. H.; Morel, J. L. and Oudot, J. (2000).Vertical infiltration of fuel oil hydrocarbons in an agricultural soil. *Toxicological and Environmental Chemistry*, 74, 111-124.

Chen, K. F.; Kao, C. M.; Chen, C. W.; Surampalli, R. Y., and Lee, M. S. (2010). Control of petroleum-hydrocarbon contaminated groundwater by intrinsic and enhanced bioremediation. *Journal of Environmental Sciences,* 22, 864-871.

Cole, J.; Beare, D. M.; Waugh, A. P.; Capulas, E.; Aldridge, K. E.; Arlett, C. F., and Farmer, P. B. (1997). Biomonitoring of possible human exposure to environmental genotoxic chemicals: lessons from a study following the wreck of the oil tanker Braer. *Environmental and molecular mutagenesis,*30, 97-111.

Cotou, E.; Castritsi-Catharios, I. and Moraitou-Apostolopoulou, M. (2001) Surfactant-based oil dispersant toxicity to developing nauplii of *Artemia*: effects on ATPase enzymatic system. *Chemosphere*, 42, 959–964.

Cripps, G. C. and Priddle, J. (1991).Hydrocarbons in the Antarctic environment. *ANS*, 3, 233-250.

Crum, J. E. (1993). Peak expiratory flow rate in schoolchildren living close to Braer oil spill. *BMJ*, 307, 23–24.

Daane, L. L.; Harjono, I.; Zylstra, G. J., and Haggblom, M. M. (2001). Isolation and characterization of polycyclic aromatic hydrocarbon-degrading bacteria associated with the rhizosphere of salt marsh plants. *Applied and Environmental Microbiology*, 67, 2683-2691.

Danchuk, S. and Willson, C. S. (2010). Effects of shoreline sensitivity on oil spill trajectory modeling of the Lower Mississippi River. *Environmental Science and Pollution Research*, 17, 331–340.

Dave, D. and Ghaly, A. E. (2011). Remediation Technologies for Marine Oil Spills: A Critical Review and Comparative Analysis. *American Journal of Environmental Sciences,* 7, 423-440.

Doherty, V. F. and Otitoloju, A. A. (2013).Monitoring of soil and groundwater contamination following a pipeline explosion and petroleum product spillage in Ijegun, Lagos Nigeria. *Environmental monitoring and assessment,* 185, 4159-4170.

Doval, M. D.; Moroño, A.; Pazos, Y.; Lopez, A.; Madriñan, M.; Cabanas, J. M.; Maneiro, J. (2006) Monitoring dissolved aromatic hydrocarbon in Rias Baixasembayments (NW Spain) after Prestige oil spills: Relationship with hydrography. *Estuarine, Coastal and Shelf Science*, 67, 205-218.

Duffy, J. J.; Peake, E. and Mohtadi, M. F. (1980). Oil spills on land as potential sources of groundwater contamination. *Environment International*, 3, 107-120.

Farhadian, M.; Vachelard, C.; Duchez, D., and Larroche, C. (2008). In situ bioremediation of monoaromatic pollutants in groundwater: A review. *Bioresource Technology*, 99, 5296–5308.

Fu, W.; Fu, H.; Skøtt, K., and Yang, M. (2008).Modeling the spill in the Songhua River after the explosion in the petrochemical plant in Jilin. *Environmental Science and Pollution Research*, 15, 178–181.

Gargouri, B.; Karray, F.; Mhiri, N.; Aloui, F., and Sayadi, S. (2013). Bioremediation of petroleum hydrocarbons-contaminated soil by bacterial consortium isolated from an industrial wastewater treatment plant. *Journal of Chemical Technology and Biotechnology*, 89, 978-987.

Garoma, T.; Gurol, M. D.; Osibodu, O.,andThotakura, L. (2008). Treatment of groundwater contaminated with gasoline components by an ozone/UV process.*Chemosphere*,73, 825-831.

Gerhardt, K. E.; Huang, X. D.; Glick, B. R., and Greenberg, B. M. (2009). Phytoremediation and rhizoremediation of organic soil contaminants: potential and challenges. *Plant Science*, 176, 20-30.

Gómez Gesteira, J. L. And Dauvin, J. C. (2000). Amphipods are good bioindicators of the impact of oil spills on soft bottom macrobenthos communities. *Marine Pollution Bulletin*, 40, 1017–1027.

Gomez Gesteira, J. L.; Dauvin, J. C.,and Salvande Fraga, M. (2003).Taxonomic level for assessing oil spill effects on soft-bottom sublittoral benthic communities. *Marine Pollution Bulletin*, 46, 562–572.

Gomez, F. and Sartaj, M. (2013). Field scale ex-situ bioremediation of petroleum contaminated soil under cold climate conditions. *International Biodeterioration and Biodegradation*, 85, 375-382.

González, J. J.;Viñas, L.; Franco, M. A.; Fumega, J.; Soriano, J. A.; Grueiro, G.; Muniategui, S.; López-Mahía, P.; Prada, D.; Bayona, J. M.; Alzaga, R.,andAlbaigés, J. (2006). Spatial and temporal distribution of dissolved/dispersed aromatic hydrocarbons in seawater in the area affected by the Prestige oil spill. *Marine Pollution Bulletin*, 53, 250–259.

Guerin, T. F.; Horner, S.; McGovern, T., and Davey, B. (2002). An application of permeable reactive barrier technology to petroleum hydrocarbon contaminated groundwater. *Water Research*, 36, 15-24.

Gundlach, E. R.; Boehm, P. D.; March and, M.; Atlas, R. M.; Ward, D. M., and Wolfe, D. A. (1983) The fate of Amoco Cadiz oil. *Science*, 221, 122-129.

Hamamura, N.; Fukui, M.; Ward, D. M., and Inskeep, W. P. (2008).Assessing soil microbial populations responding to crude-oil amendment at different temperatures using phylogenetic, functional gene (alkB) and physiological analyses. *Environmental science and technology*,42, 7580-7586.

Harvey, J. S.; Lyons, B. P.; Page, T. S.; Stewart, C., and Parry, J. M. (1999).An assessment of the genotoxic impact of the Sea Empress oil spill by the measurement of DNA adduct levels in selected invertebrate and vertebrate species. *Mutation Research/Genetic Toxicology and Environmental Mutagenesis,* 441, 103-114.

Hassanshahian, M.; Tebyanian, H. and Cappello, S. (2012). Isolation and characterization of two crude oil-degrading yeast strains, *Yarrowialipolytica* PG-20 and PG-32, from the Persian Gulf. *Marine pollution bulletin*, 64, 1386-1391.

Hayworth, J. S. and Clement, T. P. (2011). BP's Operation Deep Clean—could dilution be the solution to beach pollution? *Environmental Science and Technology,* 45, 4201–4202.

Hazen, T. C.; Dubinsky, E. A.; Sesantis, T. Z.; Andersen, G. L., et al. (2010). Deep-sea oil plume enriches indigenous oil-degrading bacteria. *Science*, 330, 204-208.

Hemmer, M. J.; Barron, M. G. and Greene, R. M. (2011). Comparative toxicity of eight oil dispersants, Louisiana sweet crude oil (LSC), and chemically dispersed LSC to two aquatic test species. *Environmental Toxicology and Chemistry*, 30, 2244–2252.

Holakoo, L.(2001). *On the capability of rhamnolipids for oil spill control of surface water.* Concordia University.

Hong, S.; Khim, J. S.; Ryu, J.; Park, J.; Song, S. J.; Kwon, B. O.; Choi, K.; Ji, K.; Seo, J.; Lee, S.; Park, J.; Lee, W.; Choi, Y.; Lee, K. T.; Kim, C. K.; Shim, W. J.; Naile, J. E., and Giesy, J. P. (2012). Two years after the Hebei Spirit oil spill: residual crude-derived hydrocarbons and potential AhR-mediated activities in coastal sediments. *Environmental Science and Technology*, 46, 1406–1414.

Hosmer, A. W.; Stanton, C. E. and Beane, J. L. (1997). Intent to Spill: Environmental effects of Oil Spills Caused by War, Terrorism, Vandalism, and Theft. In: *International Oil Spill Conference*. American Petroleum Institute, 157-163.

Huesemann, M. H.; Hausmann, T. S. and Fortman, T. J. (2004). Does bioavailability limit biodegradation? A comparison of hydrocarbon biodegradation and desorption rates in aged soils. *Biodegradation*, 15, 261-274.

Ijah, U. J. J. (1998).Studies on relative capabilities of bacterial and yeast isolates from tropical soil in degrading crude oil. *Waste Management*., 18, 293–299.

Incardona, J. P.; Gardner, L. D.; Linbo, T. L.; Brown, T. L.; Esbaugh, A. J.; Mager, E. M., and Scholz, N. L. (2014). Deepwater Horizon crude oil impacts the developing hearts of large predatory pelagic fish. *Proceedings of the National Academy of Sciences*, 111, E1510-E1518.

InformeTécnico CSIC-Prestige-13.*Caracterización de lafraccionsolubleenel agua delfuel-oil vertido por el Prestige. (2003).*Http://csicprestige.iim. Csic.es/desarro/informcsic/13 /index.htm (accessed 26.04.14).

Irons, D. B.; Kendall, S. J.; Erickson, W. P.; Mcdonald, L. L., and Lance, B. K. (2000). Nine years after the Exxon Valdezoil spill: Effects on marine bird populations in Prince William Sound, Alaska. *Condor*, 102, 723–737.

James, C. A. and Strand, S. E. Phytoremediation of small organic contaminants using transgenic plants. *Current Opinion in Biotechnology*, 20, 237-241, 2009.

Janjua, N. Z.; Kadir, M. M.; Lutfi, S.; Tipre, M., and Sathiakumar, N. (2013). Tasman spirit oil spill in Pakistan: Research response and lessons learned. *American journal of industrial medicine*, 56, 124-131.

Jernelöv, A. (2010). The threats from oil spills: Now, then, and in the future. *Ambio*, 39, 353-366.

Ji, K.; Seo, J.; Liu, X.; Lee, J.; Lee, S.; Lee, W.; Park, J.; Khim, J. S.; Hong, S.; Choi, Y.; Shim, W. J.; Takeda, S.; Giesy, J. P., and Choi, K. (2011). Genotoxicity and endocrine-disruption potentials of sediment near an oil spill site: two years after the Hebei Spirit oil spill. *Environmental Science and Technology*, 45, 7481–7488.

Joergensen, R. G.; Figge, R. M. and Kupsch, L. (1997).Microbial decomposition of fuel oil after compost addition to soil. *Zeitschriftfür Pflanzenernährung und Bodenkunde*, 160, 21-24.

Jung, J. H.; Kim, M.; Yim, U. H.; Ha, S. Y.; An, J. G.; Won, J. H.; Han, G. M.; Kim, N. S.; Addison, R. F., and Shim, W. J. (2011). Biomarker responses in pelagic and benthic fish over 1 year following the Hebei Spiritoil spill (Taean, Korea).*Marine Pollution Bulletin*, 62, 1859–1866.

Jung, S. C.; Kim, K. M.; Lee, K. S.; Roh, S.; Jeong, W. C.; Kwak, S. J., and Jee, Y. K. (2013). Respiratory Effects of the Hebei Spirit Oil Spill on Children in Taean, Korea. *Allergy, asthma and immunology research*, 5, 365-370.

Junoy, J.; Castellanos, C.; Vieitez, J. M.; LaHuz, M. R.,andLastra, M. (2005). The macroinfauna of the Galician sandy beaches (NW Spain) affected by the Prestige oil-spill. *Marine Pollution Bulletin*, 50, 526–536.

Kim, M.; Hong, S. H.; Won, J.; Yim, U. H.; Jung, J. H.; Ha, S. Y.; An, J. G.; Joo, C.; Kim, E.; Han, G. M.; Baek, S.; Choi, H. W., and Shim, W. J. (2013). Petroleum hydrocarbon contaminations in the intertidal seawater after the Hebei Spirit oil spill – Effect of tidal cycle on the TPH concentrations and the chromatographic characterization of seawater extracts. *Water Research*, 47, 758–768.

Kim, M.; Yim, U. H.; Hong, S. H.; Jung, J. H.; Choi, H. W.; An, J.; Shim, W. J., and Won, J. (2010). Hebei Spirit oil spill monitored on site by fluorometric detection of residual oil in coastal waters off Taean. Korea. *Marine pollution bulletin*, 60, 383-389.

Kingston, P. (1999). Recovery of the Marine Environment Following the *Braer* Spill, Shetland. *International Oil Spill Conference Proceedings*, 1999, 103-109.

Kingston, P. F. (2002). Long-term environmental impact of oil spills. *Spill Science and Technology Bulletin*, 7, 53–61.

Kingston, P. F.; Dixon, I. M. T.; Hamilton, S., and Moore, D. C. (1995). The impact of the Braer oil-spill on the macrobenthicinfauna of the sediments of the Shetland islands. *Marine Pollution Bulletin* 30, 445–459.

Ko, J. Y. and Day, J. W. (2004). A review of ecological impacts of oil and gas development on coastal ecosystems in the Mississipi Delta. *Ocean and Coastal Management*, 47, 597-623.

Kostka, J. E.; Prakash, O.; Overholt, W. A.; Green, S. J.; Freyer, G.; Canion, A.; Delgardio, J.; Norton, N.; Hazen, T. C.,and Huettel, M. (2011).Hydrocarbon-degrading bacteria and the bacterial community response in Gulf of Mexico beach sands impacted by the Deepwater Horizon oil spill. *Applied and environmental microbiology,* 77, 7962-7974.

Krahn, M. M.; Kittle, L. J. J. and Macleod, W. D. J. (1986) Evidence for exposure of fish to oil spilled into the Columbia River. *Marine Environmental Research,* 20, 291–298.

Laffon, B.; Aguilera, F.; Ríos-Vázquez, J.; Valdiglesias, V.,andPásaro, E. (2014). Follow-up study of genotoxic effects in individuals exposed to oil from the tanker Prestige, seven years after the accident. *Mutation Research,* 760, 10–16.

Laffon, B.; Fraga-Iriso, R.; Pérez-Cadahía, B., and Méndez, J. (2006).Genotoxicity associated to exposure to Prestige oil during autopsies and cleaning of oil-contaminated birds. *Food Chemical Toxicology*, 44, 1714–1723.

LeHir, M. and Hily, C. (2002) First observations in a high rocky-shore community after the Erika oil spill. *Marine pollution bulletin*, 44, 1243-1252.

Leme, D. M.; Angelis, D. F. and Marin-Morales, M. A. (2008). Action mechanisms of petroleum hydrocarbons present in waters impacted by an oil spill on the genetic material of Allium cepa root cells. *Aquatic Toxicology,* 88, 214–219.

Lemiere, S.; Cossu-Leguille, C.; Bispo, A ; Jourdain, M. J.; Lanhers, M. C.; Burnel, D., and Vasseur, P. (2005). DNA damage measured by the single-cell gel electrophoresis (Comet) assay in mammals fed with mussels contaminated by the 'Erika'oil-spill. *Mutation Research/Genetic Toxicology and Environmental Mutagenesis,* 581, 11-21.

Lessard, R. R. and Demarco, G. (2000).The significance of oil spill dispersants. *Spill Science and Technology Bulletin*, 6, 59-68.

Lindén, O. and Pålsson, J. (2013).Oil contamination in Ogoniland, Niger Delta. *AMBIO*, 42, 685–701.

Lopes, P. F. (2010) *Diversidade taxonômica e potencial de biodegradação de bactérias isoladas de reservatórios de petróleo da Bacia de Campos (RJ).* Universidade de São Paulo

López, E.;Schuhmacher, M.; Domingo, J. L. (2008). Human health risks of petroleum-contaminated groundwater. *Environmental Science and Pollution Research,* 15, 278–288.

Lyons, B. P.; Harvey, J. S. and Parry, J. M. (1997).An initial assessment of the genotoxic impact of the Sea Empress oil spill by the measurement of DNA adduct levels in the intertidal teleost Lipophryspholis. *Mutation Research,* 390, 263–268.

Lyons, R. A.; Temple, J. M. F.; Evans, D.; Fone, D. L., and Palmer, S. R. (1999). Acute health effects of the Sea Empress oil spill. *Jounal Epidemiol. Community Health,* 53, 306–310.

Maciel-Souza, M. D. C.; Macrae, A.; Volpon, A. G. T.; Ferreira, P. S.,andMendonça-Hagler, L. C. (2006). Chemical and microbiological characterization of mangrove sediments after a large oil-spill in Guanabara Bay-RJ-Brazil. *Brazilian Journal of Microbiology*, 37, 262-266.

Manee, P. Prayad, P.; Edward, S.; Upatham, A., and Ladda, T. (1998).Biodegradation of crude oil by soil microorganisms in the tropic. *Biodegradation*, 9, 83–90.

Mascolo, G.; Ciannarella, R.; Balest, L., and Lopez, A. (2008). Effectiveness of UV-based advanced oxidation processes for the remediation of hydrocarbon pollution in the groundwater: A laboratory investigation. *Journal of Hazardous Materials*, 152, 1138–1145.

Matthew, O.; Ilori, S. A.; Adebusoye, A., and Ojo, C. (2008). Isolation and characterization of hydrocarbon-degrading and biosurfactant-producing yeast strains obtained from a polluted lagoon water. *World Journal of Microbiology Biotechnology,* 24, 2539–2545.

Mazzeo, D. E. C.; Levy, C. E.; De Angelis, D. D. F., and Marin-Morales, M. A. (2010).BTEX biodegradation by bacteria from effluents of petroleum refinery. *Science of the total environment*,408, 4334-4340.

McCay, F. D.; Rowe, J. J., Whittier, N., Sankaranarayanan, S., and Etkin, D. S. (2004).Estimation of potential impacts and natural resource damages of oil. *Journal of hazardous materials*,107, 11-25.

Meo, S. A.; Al-Drees, A. M.; Meo, I. M. U.; Al-Saadi, M. M., and Azeem, M. A. (2008). Lung function in subjects exposed to crude oil spill into sea water. *Marine Pollution Bulletin*, 56, 88–94.

Michel, J.; Nixon, Z.; Dahlin, J.;Betenbaugh, D.; White, M.; Burton, D., and Turley, S. (2009). Recovery of interior brackish marshes seven years after the chalk point oil spill. *Marine Pollution Bulletin*, 58, 995–1006.

Mohsenzadeh, F.; Chehregani, R. A. and Mehrangiz, A. (2012).Evaluation of oil removal efficiency and enzymatic activity in some fungal strains for bioremediation of petroleum-polluted soils. *Iranian journal of environmental health science and engineering*,9, 1-8.

Mohsenzadeh, F.; Nasseri, S.; Mesdaghinia, A.; Nabizadeh, R.; Zafari, D.; Khodakaramian, G., and Chehregani, A. (2010). Phytoremediation of petroleum-polluted soils: Application of *Polygonumaviculare* and its root-associated (penetrated) fungal strains for bioremediation of petroleum-polluted soils. *Ecotoxicology and environmental safety*,73, 613-619.

Morita, A.; Kusaka, Y.; Deguchi, Y.; Moriuchi, A.; nakanaga, Y.; Iki, M.; Miyazaki, S., and Kawahara, K. (1999).http://www.sciencedirect.com/science/article/pii/ S00139351 99939798 - A5 Acute health problems among the people engaged in the cleanup of the Nakhodka oil spill. *Environmental Research*, 81, 185–194.

Ndimele, P. E. (2010).A review on the phytoremediation of Petroleum Hydrocarbon. *Pakistan Journal of Biological Science*, 13, 715-722.

Nedunuri, K. V.; Govindaraju, R. S.; Banks, M. K.; Schwab, A. P., and Chen, Z. (2000). Evaluation of phytoremediation for field-scale degradation of total petroleum hydrocarbons. *Journal of Environmental Engineering*, 126, 483-490.

NGWA (National Ground Water Association).(2014). *Groundwater supply and use,* http://wellowner.org/groundwater/groundwater-supply-use/, accessed in 02 de May de 2014.

Nikitik, C. C. S. and Robinson, A. W. (2003).Patterns in benthic populations in the Milford Haven waterway following the 'Sea Empress' oil spill with special reference to amphipods. *Marine Pollution Bulletin*, 46, 1125–1141.

NOAA. (1992). Department of Commerce, *The Kuwait oil fires-air quality data and reports, Gulf Pro-gram Office*, February 1992, Washington, DC, US.

Northcott, G. L. and Jones, K. C. (2001).Partitioning, extractability, and formation of nonextractable PAH residues in soil. 1. Compound differences in aging and sequestration. *Environmental science and technology*, 35, 1103-1110.

NRC (National Research Council). (1985). *Oil in the Sea: Inputs, Fates and Effects*. National Academy Press, Washington, DC.

Nwilo, P. C. and Badejo, O. T. (2004).*Management of oil spill dispersal along the Nigerian coastal areas.*

Pacwa-Plociniczak, M.; Plazza, G.A.; Poliwoda, A., and Piotrowska-Seget, Z. (2014).Characterization of hydrocarbon-degrading and biosurfactant-producing Pseudomonas sp. P-1 strain as a potential tool for bioremediation of petroleum-contaminated soil. *Environmental Science Pollution Research,* DOI 10.1007/s11356-014-2872-1.

Palinkas, L. A.; Peterson, J. S.; Russell, J., and Downs, M. A. (1993). Community patterns of psychiatric disorders after the Exxon Valdez oil spill. *The American Journal of Psychiatry*, 150, 1517–1523.

Palinkas, L. A.; Russell, J.; Downs, M. A., and Petterson, J. S. (1992). Ethnic differences in stress, coping and depressive symptoms after the Exxon Valdez oil spill. *The Journal of Nervous and Mental Disease*, 180, 287–295.

Paul, J. H.; Hollander, D.; Coble, P.; Daly, K. L.; Murasko, S.; English, D.; Basso, J.; Delaney, J.; McDaniel, L., and Kovach, C. W. (2013) Toxicity and Mutagenicity of Gulf of Mexico Waters During and After the Deepwater Horizon Oil Spill. *Environmetal Science and Technology*, 47, 9651−9659.

Pereira Jr., N.; Gomes, E. B. and Soriano, A. U. (2009) Séries em Biotecnologia. *Biodegradação de Hidrocarbonetos*. 1ª edição. Rio de Janeiro, 3.

Peterson, C. H. (2001). The "Exxon Valdez" oil spill in Alaska: Acute, indirect and chronic effects on the ecosystem. *Advances in Marine Biology,* 39, 1–103.

Peterson, C. H.; Kennicutt, M. C.; Green, R. H.; Montagna, P.; Harper, J. D. E.; Poweel, E. N., and Roscigno, P. F. (1996). Ecological consequences of environmental perturbations associated with offshore hydrocarbon production: a perspective from study of ling-term exposures in the Gulf of Mexico. *Canadian Journal of Fisheries and Aquatic Sciences*, 53, 2637-2654.

Peterson, C. H.; Rice, S. D.; Short, J. W.; Esler, R. D.; Bodkin, J. L.; Ballachey, B. E., and Irons, D. B. (2003). Long-term ecosystem response to the Exxon Valdez oil spill. *Science*, 302, 2082–2086.

Pradhan, S. P.; Conrad, J. R.; Paterek, J. R., and Srivastava, V. J. (1998).Potential of phytoremediation for treatment of PAHs in soil at MGP sites. *Journal of soil contamination*,7, 467-480.

Prince, R. C. and Butler, J. D. (2013).A protocol for assessing the effectiveness of oil spill dispersants in stimulating the biodegradation of oil.*Environmental Science and Pollution Research*, 1–5.

Prince, R. C. (2010) Eukaryotic hydrocarbon degraders. In: Timmis, K. N. (ed.) *Handbook of hydrocarbon and lipid microbiology*. Springer, 2066–2078.

Prince, R. C.; Gramain, A. and McGenity, T. J. (2010) Prokaryotic hydrocarbon degraders. In: Timmis, K. N. (ed.) *Handbook of hydrocarbon and lipid microbiology.* Springer, 1672–1692.

Prince, R. C.; McFarlin, K. M.; Butler, J. D.; Febbo, E. J.; Wang, F. C.,and Nedwed, T. J. (2013). The primary biodegradation of dispersed crude oil in the sea. *Chemosphere*, 90, 521–526.

Qin, G.; Gong, D. and Fan, M. Y. (2013). Bioremediation of petroleum-contaminated soil by biostimulation amended with biochar. *International Biodeterioration and Biodegradation*, 85, 150-155.

Rodríguez-Trigo, G.; Zock, J. P.; Pozo-Rodríguez, F.; Gómez, F. P.; Monyarch, G.; Bouso, L., and Barberà, J. A. (2010). Health changes in fishermen 2 years after clean-up of the Prestige oil spill. *Annals of Internal Medicine*, 153, 489-498.

Samain, J. F.; Moal, J.; Coum, A.; Le Coz, J. R., and Daniel, J. Y. (1980). Effects of the "Amoco Cadiz" oil spill on zooplankton.*HelgoländerMeeresuntersuchungen*,34, 225-235.

SEEEC (1996) *Relatório Inicial Comissão de Avaliação Ambiental Sea Empress,* SEEEC, Cardiff.

Seo, J. Y.; Kim, M.; Lim, H. S., and Choi, J. W. (2014). The macrofaunal communities in the shallow subtidal areas for the first 3 years after the Hebei Spirit oil spill. *Marine Pollution Bulletin*,82, 208-220.

Sergy, G. A. and Owens, E. H. (2011).Differences and Similarities in Freshwater and Marine Shoreline Oil Spill Response. *International Oil Spill Conference Proceedings.*

Shultz, J. M.; Walsh, L.; Garfin, D. R.; Wilson, F. E., and Neria, Y.(2014). The 2010 Deepwater Horizon Oil Spill: The Trauma Signature of an Ecological Disaster. *The Journal of Behavioral Health Services and Research*, 1-19.

Sim, M. S.; Jo, I. J. and Song, H. G. (2010). Acute health problems related to the operation mounted to clean the Hebei Spirit oil spill in Taean, Korea. *Marine Pollution Bulletin*, 60, 51–57.

Sodré, F. F.; Lenzi, E. and Costa, A. C. S. (2001).Utilização de modelos físico-químicos de adsorção no estudo do comportamento do cobre em solos argilosos.*Química Nova*, 24, 324-330.

Speight, J. G. (2014). *The chemistry and technology of petroleum: The chemistry and technology of petroleum* (5th edition) CRC press Taylor and Francis Group.

Spies, R. B. and Desmarais, D. J. (1983).Natural isotope study of trophic enrichment of marine benthic communities by petroleum. *Marine Biology*, 73, 67–71.

Suárez, B.; Lope, V.; Pérez-Gómez, B.; Aragonés, N.; Rodríguez-Artalejo, F.; Marqués, F.,andPollán, M. (2005). Acute health problems among subjects involved in the cleanup operation following the Prestige oil spill in Asturias and Cantabria (Spain).*Environmental Research*,99, 413-424.

Swedmark, M.; Granmo, Å. And Kollberg, S. (1973). Effects of oil dispersants and oil emulsions on marine animals*Water Research*, 7, 1649-1672.

Tang, J.; Lu, X.; Sun, Q., and Zhu, W. (2012).Aging effect of petroleum hydrocarbons in soil under different attenuation conditions. *Agriculture, Ecosystems and Environment,*149, 109-117.

Teas, C.; Kalligeros, F.; Zanikos, S., and Stournas, E. (2001). Investigtion. Investigation of the effectiveness of absorbent materials in oil spills clean up. *Desalination*, 140, 259-264.

Trindade, P. V. O.; Sobral, L. G.; Rizzo, A. C. L.; Leite, S. G. F.,and Soriano, A. U. (2005). Bioremediation of a weathered and a recently oil-contaminated soils from Brazil: a comparison study. *Chemosphere*, 58, 515-522.

Tyagi, M.; Da Fonseca, M. M. R. and Carvalho, C. C. C. R. (2011) Bioaugmentation and biostimulation strategies to improve the effectiveness of bioremediation processes. *Biodegradation*, 22, 231–241.

Valentine, D. L.; Kessler, J. D.; Redmond, M. C.; Mendes, S. D.; Heintz, M. B.; Farwell, C.; Hu, L.; Kinnaman, F. S.; Yvon-Lewis, S.; Du, M. R.; Chan, E. W.; Tigreros, F. G.,and Villaneva, C. J. (2010). Propane respiration jump-starts microbial response to a deep oil spill. *Science*, 330, 208−211.

VanHamme, J. D.; Singh, A. and Ward, O. P. (2003). Recent advances in petroleum microbiology. *Microbiology and Molecular Biology Reviews*, 67, 503-549.

Vidali, M. (2001). Bioremediation. An overview.*Pure and Applied Chemistry*, 73, 1163-1172.

Vilcheck, G. E. and Tishkov, A. A. (1997). Usinsk oil spill. *Disturbance and Recovery in Arctic Lands.* Kluwer Academic, Dordrecht, 411-420.

Wang, J.; Xu, H. and Guo, S. (2007). Isolation and characteristics of a microbial consortium for effectively degrading phenanthrene. *Petroleum Science,* 4, 68-75.

Wang, J.; Zhang, Z.; Su, Y.; He, W.; He, F., and Song, H. (2008). Phytoremediation of petroleum polluted soil. *Petroleum Science*,5, 167-171.

Wang, Z.; Fingas, M.; Lambert, P.; Zeng, G.; Yang, C., and Hollebone, B. (2004). Characterization and identification of the Detroit River mystery oil spill. *Journal of Chromatography A*, 1038, 201–214.

Wardley-Smith, J. (1983). The dispersant problem. *Marine Pollution Bulletin*, 14, 245-249.

Winquist, E.; Björklöf, K.; Schultz, E.; Räsänen, M.; Salonen, K.; Anasonye, F.; Cajthaml, T.; Steffen, K. T., and Jørgensen, K. S. (2014). Bioremediation of PAH-contaminated soil with fungi - From laboratory to field scale. *International Biodeterioration and Biodegradation*, 86, 238-247.

Xu, S. Y.; Chen, Y. X.; Wu, W. X.; Wang, K. X.; Lin, Q., and Liang, X. Q. (2006). Enhanced dissipation of phenanthrene and pyrene in spiked soils by combined plants cultivation. *Science of the Total Environment*, 363, 206-215.

Yeh, C. H.; Lin, C. W. and Wu, C. H. (2010). A permeable reactive barrier for the bioremediation of BTEX-contaminated groundwater: Microbial community distribution and removal efficiencies. *Journal of Hazardous Materials,* 178, 74-80.

Yim, U. H.; Kim, M.; Ha, S. Y.; Kim, S., and Shim, W. J. (2012). Oil spill environmental forensics: the Hebei Spirit oil spill case. *Environmental Science and Technology*, 46, 6431-6437.

Zenetos, A.; Hatzianestis, J.; Lantzouni, M.; Simboura, M.; Sklivagou, E., and Arvanitakis, G. (2004) The Eurobulker oil spill: mid-term changes of some ecosystem indicators. *Marine Pollution Bulletin*, 48, 122–131.

Zhu, K.; Shang, Y.; Sun, P.; Li, Z.; Li, X.; Wei, J.; Wang, K.; Wu, D.; Cao, A., and Zhu, H. (2013). Oil spill cleanup from sea water by carbon nanotube sponges. *Frontiers of Materials Science*, 7, 170–176.

Zinjarde, S. S. and Pant, A. A. (2002).Hydrocarbon degraders from tropical marine environments. *Marine pollution bulletin*, 44, 118-121.

In: Oil Spills
Editor: Adam Clifton

ISBN: 978-1-63321-515-3

Chapter 5

NEW APPROACHES IN OIL SPILL MAPPING AND EMERGENCY PLANNING

A. F. Romero**, R. F. C. Fontes, M. Oliveira and D. M. S. Abessa
UNESP, Campus Experimental do Litoral Paulista,
Núcleo de Estudos sobre Poluição e Ecotoxicologia Aquática (NEPEA).
Praça Infante Dom Henrique, São Vicente, São Paulo, Brazil

ABSTRACT

During oil spill episodes, the oil may reach the coastal environments, despite efforts to retain and remove the oil from the water. Thus, response actions must be properly planned in order to minimize the environmental impacts caused by the oil. Contingency plans for oil spill accidents were developed in the 1970s and involve a set of steps that include environmental sensitivity index maps (ESI maps). At that time, this valuable tool represented advancement in action planning. Since then, however, contingency plans have used the ESI maps as a definitive planning solution, without combining them with other tools such as numerical modeling. The technological advances involved in numerical modeling, computational and geo-processing techniques, and access to a wide database through the internet, all allow for the introduction of advancements in oil spill response action plans. This chapter aims to present some of these new techniques in order to improve the classic approaches and provide more efficiency for emergency action plans. One of these techniques is the combination of ESI maps with information from numerical modeling of oil slicks and the probability that a given environment will be reached by the oil. This combination helps to generate maps of the indices of environmental vulnerability to oil (IEVO maps), which quickly and easily show areas that should receive priority protection after an oil spill. Another planning tool that may be used with the ESI and IEVO maps is the map of cleaning procedures and response actions. This map relies on the use of pictographic icons that represent the cleaning procedures recommended for each coastal environment. A careful definition of the recommended methods may keep the response actions from amplifying the damage caused by the oil, and their representation as icons on a map also allows for rapid visualization, and thus rapid decision making by officials responsible for emergency

* Corresponding author: agatafr@gmail.com

actions. These maps must also rely on solid information; thus, the assessment of biological resources of each coastal environment needs to be complete and deep. The use of GIS techniques to spatially display the biological information may help to identify more sensitive areas within a region that has already been classified using a given environmental sensitivity index. ESI maps that prioritize the representation of biological resources along the potentially affected areas provide an advancement in terms of the protection and mitigation of impacts on coastal ecosystems. These new approaches allow for better emergency planning move the classic approach forward, thus contributing to the mitigation of potential environmental impacts caused by oil spills.

INTRODUCTION

The maritime transport of cargo around the world has increased continuously over time, and has followed the increase of global trade and commerce. The volume of transported cargo has risen from 8.4 billion tons in 2010 to almost 9 billion tons in 2011 (UNCTAD, 2011). With this increase, the global production and transport of oil and its derivatives has grown as well. More intense shipping traffic, including oil tankers, has contributed to increased risks of accidents involving oil spills.

Although the relative number of oil spills caused by vessels and tankers has dropped over time, a single oil spill can produce catastrophic damages (HÖFER, 1998; HUIJER, 2005). For example, the 1989 oil spill from the Exxon Valdez tanker released approximately 36,000 tons of crude oil and affected 1,500 km of the Alaskan coastline (VAN DE WIEL & VAN DORP, 2011).It is important to note that minor spills have frequently resulted from operational failures, whereas major spills have been the result of collisions and of the grounding of vessels (POFFO et al.; 2001; HUIJER, 2005). In addition, major oil spills may originate from offshore oil production platforms (oil rigs), and they have potential to cause environmental impacts of great magnitude. On April 20th 2010 the explosion of the Deepwater Horizon platform in the Gulf of Mexico caused the spillage of approximately 5 million barrels into the sea (RAMSEUR, 2010).

It is impossible to predict when oil spills will occur. Thus, emergency response actions should be carefully planned in order to minimize the impacts that the oil has on coastal and marine environments. Due to the worldwide risks posed by oil exploitation and transportation, governments, oil companies, ship owners and international agencies have developed and established a set of laws and regulations that include the need for the development of contingency plans. A good contingency plan provides a prompt response to an oil spill (USEPA, 1999).

According to the IPIECA (2000), an oil spill contingency plan should be comprised of three parts:

- A strategy section that describes the scope of the plan, including the geographical coverage, perceived risks, roles/responsibilities of those charged with implementing the plan, and the proposed response strategy;
- An action and operations section that lays out the emergency procedures that will allow for rapid assessment of the spill and the mobilization of appropriate response resources;

- A data directory containing all relevant maps, resource lists and data sheets required to support an oil spill response effort and conduct the response according to an agreed upon strategy.

The contingency plan can be developed at different levels or within different scopes, from small oil spills in a local area to a large oil spill on a regional or national scale. According to the International Tanker Owners Pollution Federation Limited (ITOPF) (http://www.itopf.com/spill-response/contingency-planning), three oil spill levels have been defined according to the volume of oil spilled: Tier 1 (small spills that can be dealt with locally); Tier 2 (oil spills affecting a larger area and requiring better response actions along with a local contingency plan); and Tier 3 (larger oil spills that must be attended to with regional or national contingency plans).

Despite the differences that can exist among contingency plans, it is possible to say that a contingency plan usually provides information on hazard identification, vulnerability analysis, risk assessment, and response actions (USEPA, 1999).

Some of the procedures and approaches traditionally used in oil contingency planning were developed during 1970s and have slowly incorporated the new tools that have been provided as a result of technological development. This chapter will outline some new and innovative approaches that may be employed in order to improve oil contingency planning.

Environmental Sensitivity Maps (ESI Maps)

Contingency plans incorporate tools that assist decision making, indicate the best procedures to protect the local fauna and flora, and define the more sensitive and priority areas to be protected. The ESI maps are present in contingency plans and provide information on the sensitivity of the mapped areas by indicating the presence of sensitive biological and socioeconomic resources. The use of ESI Maps during the planning stages makes it easier to identify areas that should be protected first, and thus help to organize response strategies.

The determination of the sensitivity of environments was first presented by Gundlach and Hayes (1978), who proposed classifying environments according to a sensitivity index that ranged from 1 to 10 (e.g., from the least sensitive to the most sensitive). The sensitivity index was determined based on the interaction of the shoreline with the physical processes that control the deposition of oil, the persistence of oil in the environment, and the extent of damage to wildlife. When coastal environments are classified by their sensitivity to oil, the slope, wave exposure, and substrate type are considered. Thus, sheltered environments with low slopes and unconsolidated substrate, such as mangroves, are more sensitive than exposed environments with higher slopes and consolidated substrate, such as exposed rocky shores.

Based on Gundlach and Hayes (1978), the NOAA (National Oceanic and Atmospheric Administration) has proposed a set of Environmental Sensitivity Index (ESI) Maps, which have been incorporated into U.S. contingency plans since 1979 (NOAA, 2002). These ESI maps provide information on the sensitivity of the environments using a sensitivity scale of 1 to 10, and also on the presence of both biological human-use resources (NOAA, 2002; MMA, 2004). In the early 2000s, the Brazilian Ministry of the Environment (MMA) published a handbook elaborating ESI maps based on the experience of the NOAA (MMA, 2004) (Table 1).

Table 1. Sensitivity Index for Environments (NOAA, 2002)

Sensitivity index	Environments
ESI 1A	Exposed rocky shores (estuarine/lacustrine); Exposed rocky banks (riverine)
ESI 1B	Exposed, solid, man-made structures (estuarine/lacustrine/riverine)
ESI 1C	Exposed rocky cliffs with boulder talus base (estuarine/lacustrine/ riverine)
ESI 2A	Exposed scarps and steep slopes in clay (estuarine)
ESI 2B	Exposed scarps and steep slopes in clay (estuarine)
ESI 3A	Fine to medium-grained sand beaches (estuarine)
ESI 3B	Scarps and steep slopes in sand (estuarine); Eroding scarps in unconsolidated sediment (lacustrine); Exposed, eroding banks in unconsolidated sediment (riverine)
ESI 3C	Tundra cliffs (estuarine)
ESI 4	Coarse-grained exposed sand beaches (estuarine); Sand beaches (lacustrine); Sandy bars and gently sloping banks (riverine)
ESI 5	Mixed gravel-sand beaches (estuarine/lacustrine); Mixed gravel-sand bars and gently sloping banks (riverine)
ESI 6A	Gravel beaches (estuarine/lacustrine); Gravel bars and gently sloping banks (riverine)
ESI 6B	Gravel beaches/ riprap (estuarine/lacustrine/riverine)
ESI 7	Exposed tidal flat (estuarine/lacustrine)
ESI 8A	Sheltered rocky shore (estuarine); Sheltered scarps in bedrock, mud or clay (estuarine/lacustrine)
ESI 8B	Sheltered, solid man-made structures (estuarine/lacustrine/riverine)
ESI 8C	Sheltered riprap (estuarine/lacustrine/riverine)
ESI 8D	Sheltered rocky rubble shores (estuarine)
ESI 8E	Peat shorelines (estuarine)
ESI 8F	Vegetated, steeply-sloping bluffs (riverine)
ESI 9A	Sheltered tidal flats (estuarine); Sheltered sand/mud flats (lacustrine)
ESI 9B	Vegetated low banks (estuarine/ lacustrine/ riverine)
ESI 9C	Hypersaline tidal flats (estuarine)
ESI 10A	Saltwater and brackish-water marshes (estuarine)
ESI 10B	Freshwater marshes (estuarine/ lacustrine/riverine)
ESI 10C	Swamps (estuarine/lacustrine/riverine)
ESI 10D	Scrub-shrub wetlands (estuarine/lacustrine/riverine); Mangroves (estuarine)
ESI 10E	Inundated low-lying tundra (estuarine)

Preparedness for oil spill responses is a fundamental issue for protecting marine environments from pollution. During real situations of environmental accidents, the decision-making process is very difficult due to the lack of time and resources. Moreover, managers and/or stakeholders responsible for the oil spill response actions in the field are rarely experts in environmental topics. Thus, it is very important that scientists develop a tool to support decision making, one that includes easily understood environmental information (Ihaksi et al., 2011)that is effective in real situations.

ESI Maps are a valuable tool for contingence planning; however, to better prevent or mitigate impacts caused by oil spills, it is necessary to integrate other planning tools, such as the numerical modeling of oil slicks (Michel et al., 1978).

Maps of Environmental Vulnerability to Oil

Maps of Environmental Vulnerability to Oil (MEVO) may be important tools for response actions after oil spills, because they present information on the sensitivity and susceptibility of the mapped areas. Highly sensitive areas with a high probability of being affected by oil are considered more vulnerable than areas with low sensitivity levels and a low probability of being affected. Vulnerability maps should therefore bring together information on a given environment's sensitivity to oil. This information should be obtained from the widest possible combination of environmental aspects, and should include geomorphological, chemical, and physical characteristics, as well as biological, aesthetic, and socioeconomic consequences for the environment and the human population (Castanedo et al., 2009; Mendonza-Cantú et al., 2011). In addition, MEVOs must consider Maps of Environmental Sensitivity to Oil (ESI Maps), which present information on the sensitivity of a given environment based on the type of coastline, biological resources, and socioeconomic factors.

Gundlach and Hayes (1978) proposed the first method for classifying environments according to their sensitivity to oil. In their research, they established an index of environmental vulnerability to oil based on the geomorphological characteristics of ecosystems. Since then, the terms "sensitivity" and "vulnerability" have been used interchangeably in many studies, although they may have different meanings. A discussion of the use of these terms was presented by Silva et al. (2012). Environmental sensitivity involves how the environment responds negatively to the impact of an oil spill (Silva et al., 2012). Meanwhile, environmental vulnerability is determined by the environmental susceptibility to a certain impact, and it considers structural weakness, sensitivity, and maturity of the ecosystems involved. Environmental susceptibility to oil is the probability of a specific area being reached and affected by oil, and it depends on climatic and oceanographic conditions, spill location, and the type and amount of oil spilled. In summary, it is possible to define environmental vulnerability to oil as the interaction between the sensitivity of a given environment and the possibility of its exposure to oil (Figure 1).

Vulnerability	=	Sensitivity	x	Exposure to oil

Figure 1. Simplified definition of environmental vulnerability to oil.

International recommendations mention Maps of Environmental Vulnerability to Oil (MEVO), which should be drawn after considering the probability that the oil will reach a given area and determined through both simulations based on the worst-case scenario of oil spill and observations of previous oil spill incidents. No other information about the method for the preparation of vulnerability maps has been described. In any case, the different approaches have not been brought together as a single integrated tool that would establish priority areas in cases of oil spills, and their separated use may be dubious or conflicting, especially if ESI maps and dispersion models do not agree in their indications of priority areas to be protected from oil (Figure 2). In addition, ESI maps have been used as if they were MEVOs, and results of oil dispersion models are often ignored. Recently, the ITOPF released a set of technical papers on its webpage (www.itopf.com), discussing the problems involved in oil spills. They included contingency plans, but they failed to recommend the integration of ESI maps and dispersion models.

The use of numerical modeling determines how an environment can be exposed to oil, and particularly its susceptibility to an oil spill. The vulnerability of an area depends on: (a) the type of oil; (b) the volume spilled; (c) oceanographic and meteorological characteristics; (d) the spill point; and (e) environmental sensitivity. This chapter presents the example given by Romero et al. (2013) for the Port of Santos, in which MEVOs were developed for winter and summer oil spill scenarios.

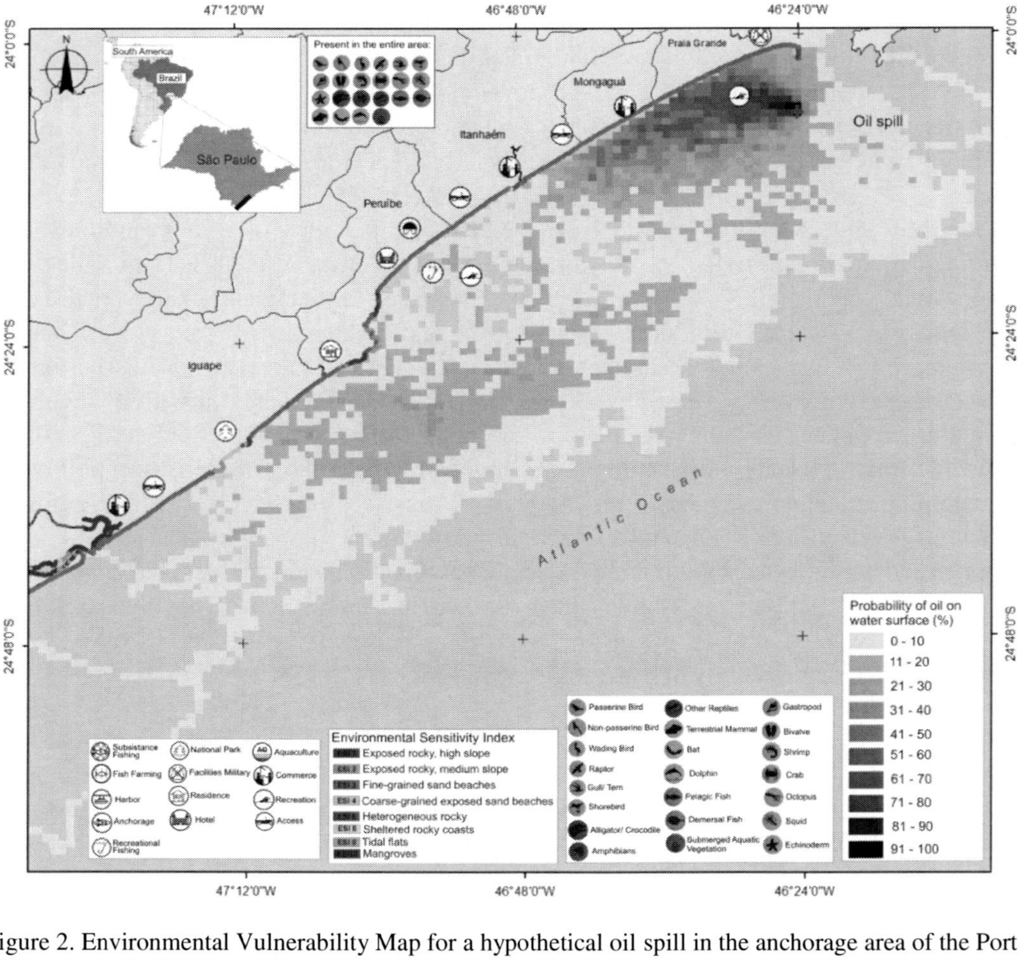

Figure 2. Environmental Vulnerability Map for a hypothetical oil spill in the anchorage area of the Port of Santos, on the São Paulo Coast, in the Santos Basin, Brazil – summer scenario (adapted from Romero et al., 2013).

The approach used in the preparation of vulnerability maps integrates the use of ESI Maps and numerical modeling of oil slick dispersion and transport. After the integration of these tools, the environment can be classified according to the index of environmental vulnerability to oil, and maps of environmental vulnerability to oil can then be generated.

In order to allow for the creation of a matrix crossing the Environmental Sensitivity Index (ESI) with the probabilities of oil reaching given coastal environments, positive and negative values should be attributed to both categories. ESIs representing low sensitivity received the code (-), those representing medium sensitivity received the code (+/-), those with high sensitivity received the code (+), and ESIs with very high sensitivity were coded (++). In the same way, codes were attributed to probabilities: low (-), medium (+/-), high (+), and very high (++) (Table 2). The combination of sensitivity and probability using the codes defined above allowed for the classification of vulnerability as low, medium, high and very high, originating the Index of Environmental Vulnerability to Oil (IEVO). Thus, for example, the IEVO 2 (low vulnerability), is determined by crossing the ESI 1 and ESI 2 (-) with the probabilities between 1% and 20% (-) (Table 3). Table 4 presents the IEVO classification according to the criteria adopted in the study. The regions where the oil does not reach the shore (probability 0%) are classified as IEVO 1 (damage not measurable).

To determine the risks to environments near the spill site, multiple scenarios and information on current wind conditions are needed. Thus, the model was run in probabilistic mode, and it generated contours of probabilities on the water surface and shoreline. The spill point is located in the anchorage area of the Port of Santos, (coordinates 24°06'04"S and 46°24'[illegible]1" W), since this area presents intense traffic of vessels, and therefore experiences greater risk of collisions with oil spills at sea. The model was run in probabilistic mode for a period of 30 days. It considered an instantaneous subsurface spill, which was defined as oil spilled to a depth of 1 m. A horizontal dispersion of 5 m^2/s, time intervals of 10 min, and a minimum thickness on the surface of 0.0003 mm, were all imagined. More details of simulation with the oil model are described in Romero et al. (2013).

The study generated scenarios that represented the trajectory of the oil slick on the water surface and the process of the oil reaching the shore. The scenarios generated are representative of summer conditions (February through April 2005). The model was run based on a volume of oil released by a Worst-Case Discharge (WCD). To calculate the WCD in this case study, the capacity of a cargo ship tank was adopted as the example. The duration of the spill was instantaneous, and the entire load was thought to be released into the ocean at once, a process which simulated a collision followed by a total hull rupture. The selected oil was MF-380 (Marine Fuel), the main fuel used in commercial and cargo ships. MF-380 is considered heavy and persistent oil that is highly toxic with low solubility in water. Figure 3 shows a plan for how vulnerability maps are generated by integrating the information from the ESI Maps and the results of the model.

Table 2. ESI and probabilities code attribution

ESI	Oil spill Probability (%)	Code	Description
1	1 to 10	-	Low sensitivity or low probability
2	11 to 20		
3	21 to 30	+ / -	Medium sensitivity or medium probability
4	31 to 40	+	High sensitivity or high probability
5	41 to 50		
6	51 to 60		
7	61 to 70	+ +	Very high sensitivity or very high probability
8	71 to 80		
9	81 to 90		
10	91 to 100		

Table 3. Results from crossing the codes used to generate the Index of Environmental Sensitivity to Oil (IEVO)

Crossing	Vulnerability	IEVO
(-) (-)	Low	2
(-) (+/-) (-) (+) (+/-) (+/-)	Medium	3
(+/-) (+) (+) (+) (+/-) (++) (-) (++)	High	4
(++) (++) (+) (++)	Very High	5

Table 4. Index Environmental Vulnerability to Oil (IEVO). IEVO 1– damage not measurable; IEVO 2 – low vulnerability; IEVO 3 – medium vulnerability; IEVO 4 – high vulnerability; and IEVO 5 – very high vulnerability

Probability (%)											
ESI	0	1-10	11-20	21-30	31-40	41-50	51-60	61-70	71-80	81-90	91-100
1	1	2	2	3	4	4	4	4	4	4	4
2	1	2	2	3	4	4	4	4	4	4	4
3	1	3	3	3	4	4	4	5	5	5	5
4	1	3	3	4	4	4	4	5	5	5	5
5	1	3	3	4	4	4	4	5	5	5	5
6	1	3	3	4	4	4	4	5	5	5	5
7	1	4	4	4	5	5	5	5	5	5	5
8	1	4	4	4	5	5	5	5	5	5	5
9	1	4	4	4	5	5	5	5	5	5	5
10	1	4	4	4	5	5	5	5	5	5	5
Index of Environmental Vulnerability to Oil – IEVO											

During the summer period, the maximum extension of coastline that can be reached by the oil slick is approximately 150 km. According to the IEVO classification, medium vulnerability (IEVO 3) was found in approximately 48% of the area considered in they hypothetical scenario for the Port of Santos, whereas no damage or not measurable effects (IEVO 1) occurred in 30% of the area, low vulnerability (IEVO 2) accounted for approximately 12%, and high vulnerability (IEVO 4) corresponded to approximately 8% of the coastline (Figure 4).

The IEVO scale proposed in the hypothetical scenario separated the areas with probability levels equal to 0% (IEVO 1 – no damage or not measurable effect) from the others. The possibility of classifying a given area as presenting “damage not measurable” (IEVO 1) allowed for a clearer visualization of areas that were not likely to be reached by oil.

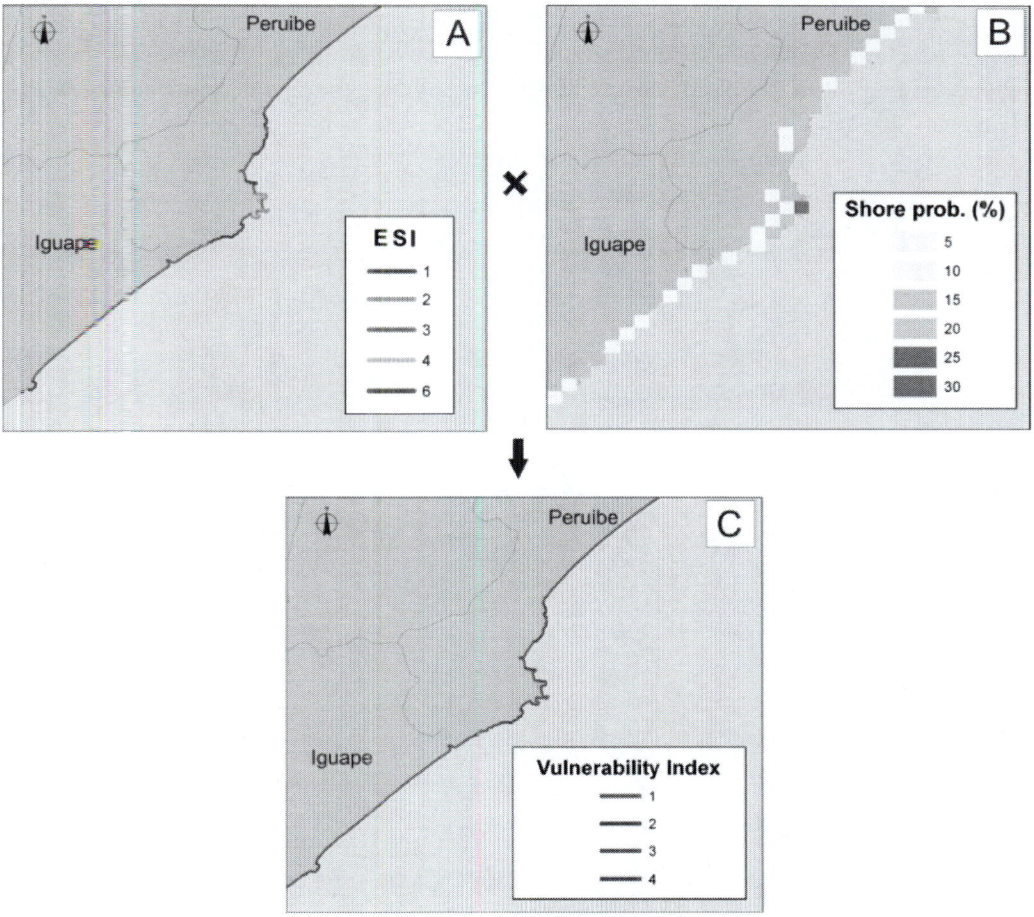

Figure 3. Plan representing the information used for the elaboration of maps of environmental vulnerability to oil using the Index of Environmental Vulnerability to Oil: (A) Environmental Sensitivity Index (ESI Maps); (B) probability of reaching the shore (numeric modeling); and (C) Index of Environmental Vulnerability to Oil (Vulnerability Map - IEVO).

This information can help planners and oil spill response teams to direct their efforts to the areas that present higher chances of being hit, thus increasing the efficiency of field actions. An efficient oil spill response may prevent the oil from reaching organisms of the mesolittoral and supralittoral zones (Romero et al., 2011), since these species are the first to suffer from the impacts of an oil slick that has reached the coast (Concawe, 1998; Geyer, 1980; Linden et al., 1979; Thomas, 1978,1973).

In the study by Romero et al. (2013), less sensitive areas (ESI 3 – exposed dissipative beaches composed of fine to medium-grained sands; exposed dune fields) were considered to be very highly vulnerable if they presented a 62% probability or higher of being reached by oil. It is important to classify relatively less sensitive environments as highly vulnerable when the communities present in those environments may be severely impacted. During the accident with the Deepwater Horizon rig in the Gulf of Mexico in 2010, the beaches along the coast were strongly hit by oil, which produced severe environmental and economic impacts (Kostka et al., 2011). A year after the accident, some beaches in the region still presented a high concentration of poly-aromatic hydrocarbons (PAH), and oil pollution was still visible on some beaches. The high PAH concentrations one year after the accident indicate that part of the remaining oil may still be causing damage to both the environment and human health (Allan et al., 2012).

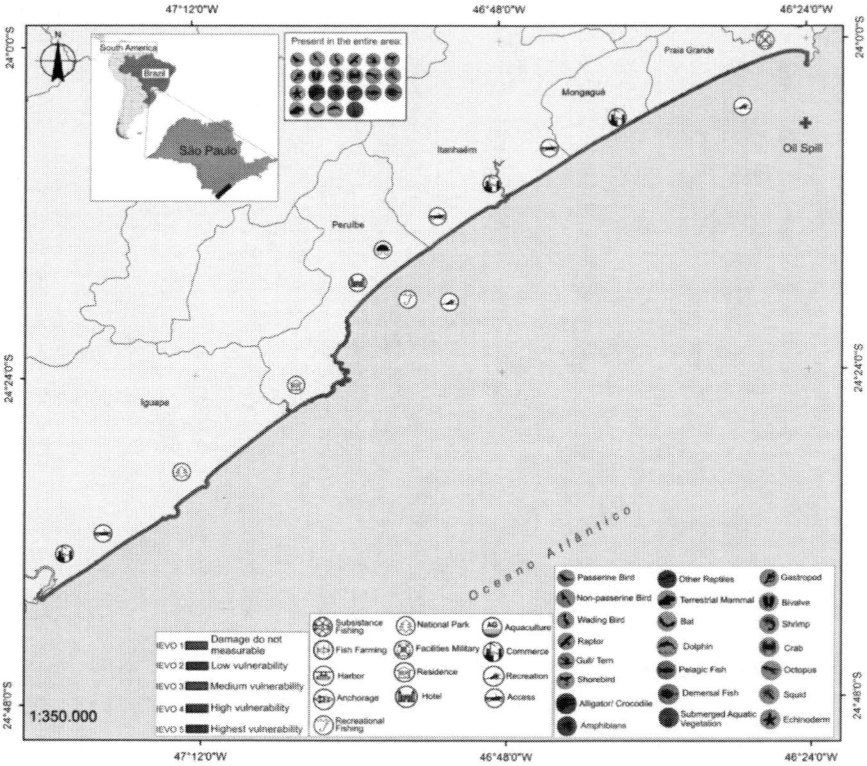

Figure 4. Environmental Vulnerability Map for a hypothetical oil spill in the anchoring area of the Port of Santos, on the São Paulo Coast, in the Santos Basin, Brazil – summer scenario (Romero et al., 2013).

In Spain, the accident with the ship Prestige in 2002 left many exposed beaches completely covered by oil. Six months after the accident, the macro-fauna of the beaches hit by oil were found to have decreased in abundance. This decrease included the elimination of some rare species, possibly due to the toxic effect of the oil and indirect effects of the cleaning procedures (Junoy et al., 2005). Thus, the vulnerability classification currently used worldwide is oversimplified and tends to group together environments with significant differences in their sensitivity to oil. In addition, it does not properly consider oil slick dispersion, and consequently, it does not provide reliable information on environmental protection. For example, smooth rocky shores (ESI 1 and 2) are grouped together with sandy beaches (ESI 3). Oil reaching those environments will behave in different manners: oil can be easily removed from smooth rocky shores through natural wind and wave actions (Muler et al., 2011; Lopes et al., 2007); however, oil may penetrate unconsolidated substrates and reach deeper layers, making its removal and decomposition more difficult. Thus, the environment may be exposed again during periods of beach erosion (Lopes et al., 2007) through trampling, storm action and bioturbation.

The development of an Index of Environmental Vulnerability to Oil (IEVO) based on the combination of data on environmental sensitivity to oil and dispersion modeling and the calculation of the probability of oil reaching the coast is a more precise and conservative approach to support the ontingency planning. This combination provides more reliable information to guide response measures during oil spills, because the data provides a quick and simplified visualization of the most vulnerable areas. The use of such indices allows the decision-making process to be more effective, and it prioritizes environments that are vulnerable according to specific conditions. In addition, maps that represent different conditions (for example, summer and winter periods) clearly show that different characteristics of oceanographic and meteorological conditions of each region in question must be considered when estimating environmental vulnerability. Thus, the proposed approach constitutes a more effective tool for the protection of coastal environments in cases of oil spills.

Maps of Cleanup Procedures and Emergency Actions

Contingency plans for oil spills have focused on providing strategies for containing oil slicks, and historically have not provided enough information for the planning of cleanup procedures in affected coastal areas.

In an oil spill accident, despite the efforts of response teams and the existence of contingency plans, coastal environments can be oiled. Oil spills require measures that will clean up the environment. The use of specific equipment and response procedures can increase the efficiency of the cleanup process in affected environments (Ventikos et al., 2004). These procedures should be chosen carefully and always with knowledge from relevant environmental agencies in order to avoid amplifying the impacts caused by oil spill (Ventikos et al., 2004). The use of high pressure jets over a rocky shore, for example, can remove not only oil, but also the organisms that live in this environment, and can therefore increase the impacts originally caused by the oil pollution.

Therefore, the selection of appropriate cleanup procedures can prevent a worsening of the oil spill impacts on the environment. In this sense, the representation of the cleanup procedures and responses to oil spills through icons on a map allows for rapid visualization and thus rapid decision-making from managers during emergency response processes.

The maps of cleanup procedures and responses to oil spills were proposed by Romero and Abessa (2014, in press) and allow for the specialization and identification of the cleanup procedures that are recommended for each environment while the procedures are still in the planning stages. The maps of cleanup procedures and response to oil spills may be prepared based on ESI Maps or MEVOs, and icons can be inserted to represent the main cleanup procedures and response actions. Romero and Abessa (2014 in press) have offered a case study in which the map of cleanup procedures and responses to oil spills was based on the ESI map and the cleanup procedures used in Brazil (represented in Table 5). However, there are other procedures used in other countries that are not recommended or approved for use in Brazil, such as in situ burning and the use of binders; thus, the icons representing these procedures were not included in Table 5.

Table 5. Icons Representing the Main Cleanup Procedures (Romero and Abessa, 2014 in press)

Icon	Description	Icon	Description
	Synthetic absorbent		Natural cleaning
	Organic absorbent in bulk		Manual cleaning
	Inorganic absorbent		Trenches
	Careful cutting of vegetation		Dispersant
	Low pressure water jet		Vacuum pumping
	High pressure water jet		Bioremediation
	Flowing water		Belt skimmer
	Sediment removal		Boom
	Skimmer		

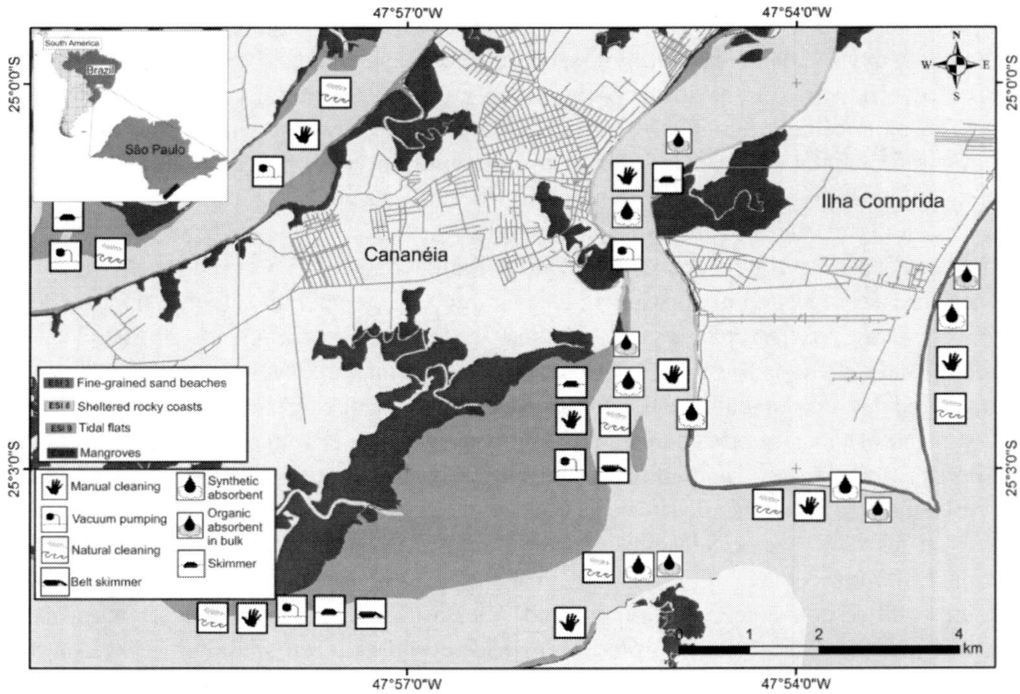

Figure 5. Map of possible cleanup procedures and response actions in cases of oil spills (from Romero and Abessa, 2014 in press).

This section presents an example of a map of cleanup procedures and responses to oil spills along a stretch of the southern coast of São Paulo state that includes the cities of Cananéia and Ilha Comprida (Figure 5). The mapped area is of high ecological and social importance, because it contains some important protected areas (Romero et al., 2010). In the region of Cananéia and Ilha Comprida, many sambaqui moundbuilders are found. They are formed by deposits of shells, stone artifacts, bones, and other remains of human activity. They represent important records of prehistoric man's activity (Diegues, 1996) and, consequently, are relevant archeological sites.

The area was characterized by the predominance of environments of high sensitivity (ESI 10 and 9), particularly mangroves and mud flats. Trampling is one of the additional impacts on mangroves and mudflats that can be caused by poorly planned responses. Trampling may kill seedlings and promote the penetration of oil into deeper layers of the substrate, where the oil will degrade more slowly (Pezeshki et al., 2000). According to Cantagallo et al. (2007) and Lopes et al. (2006), the most appropriate cleanup procedures for these environments are manual cleanup, natural cleanup, vacuum pumping, and the use of skimmers, belt harvesters, bulk organic absorbents (peat), and synthetic absorbent booms. Cleanup procedures should be restricted to areas adjacent to mangroves and mud flats where access is possible with small boats. Natural oil-absorbent pads and booms used during emergency actions must be collected after the end of operations. According to Kiesling et al. (1988), priority should be given to natural cleanup in mangroves located in areas with large tidal variation, because the action of the tides facilitates the natural removal of pollutants.

The rocky shores in the study area were classified as ESI 8 because they are situated in sheltered zones (with low hydrodynamics). In these environments, other cleanup strategies can be adopted, including the use of manual cleanup, natural cleanup, vacuum pumping, and skimmers. In tide pools close to rocky shores, the recommended cleanup approaches include the use of bulk organic absorbents (peat), though it is important not to cover organisms that can be suffocated by the absorbent (Lopes et al., 2006). Absorbent booms should be used in the places where peat was used to make its removal easier (Lopes et al., 2006). The beach areas of the studied area (on Ilha Comprida) were characterized as dissipative and exposed to wave action and present fine to medium sands; thus, they were classified as ESI 3 (Romero et al., 2010). In such environments, the use of heavy machinery, such as tractors and trucks, should be avoided, as they may cause damage to benthic organisms. Also, the removal of the oil that has reached the mesolittoral should be done carefully (manual cleanup), and only the layer of sediment that is effectively contaminated by oil should be removed. The use of synthetic absorbent booms is recommended for the subtidal region in order to prevent oil from reaching the supratidal and intertidal regions. Absorbent booms must be collected once response actions have been completed. Since beaches are exposed to waves, natural cleanup can be rather efficient in removing part of the oil that is found in subtidal and intertidal regions. The map of cleanup procedures and actions to combat oil spills must be accompanied by ESI Maps and Maps of Environmental Vulnerability to Oil (NOAA, 2002; Brazil, 2004), which provide additional information for the planning of responses to oil spills. Thus, managers and stakeholders would have an easy tool to consult in cases of oil accidents, and they could be used to organize measures quickly.

ENVIRONMENTAL AND BIOLOGICAL SENSITIVITY TO OIL SPILLS

According to Gundlach and Hayes (1978), it is difficult to predict the impact that an oil spill can have on organisms: a number of factors, such as type and volume of oil spilled, time of year, stage of life, and time of exposure to oil must be considered. The sensitive biological resources are critical information on ESI Maps, and are extremely important in determining the true sensitivity of the mapped area. In spite of the constant advancements in the methods for diagnosing human and biological resources, they are not always properly represented on ESI Maps. ESI mapping has been often negligent, showing only some groups of animals present in the area because of incomplete literature. This section will describe a proposal for elaborating a map of sensitivity based primarily on biological resources. The representation of the groups of organisms found in the estuary of Santos - São Vicente on the São Paulo coast in Brazil is used as an example (Figure 6).

This estuarine system has approximate area of 160km² and communicates with the Bay of Santos and also with the sea. The coastal ecosystems in this region are dominated by mangroves and salt marshes (classified as ESI 10), and also by rocky reefs, sand and mud flats, sheltered and semi-sheltered sandy beaches, among other types of habitats.

The surrounding area is under intense anthropic influence, including the presence of large factories, urban areas with diffuse and point sources of sewage, and the Port of Santos, the largest port in Latin America. Due to these factors, part of the studied area is already considered polluted (Afonso, 2001). The study area was divided into three sections, which are shown in Figure 6: a) the Piaçaguera Channel, b) the Port Region; c) and the Largo do Caneu section.

To produce the maps, information on biological resources was obtained from the literature (secondary data), including technical reports, books, scientific papers, academic studies (theses and dissertations) and other documents that contained information on the species. Field trips to confirm the results were also conducted.

The data was compiled and inserted into a database, where information was divided into two main areas: 1) information on the article (author, date of publication, source title, type of publication, journal name and the available data); and 2) information on groups of organisms (scientific and common name, date of collection, legal protection, ecological and socioeconomic importance, nature of the site (including indications if the data on the organism is geo-referenced or not, and the level of taxonomic identification; e.g., if identification reached the species level).This information was then entered into the GIS. The limits (start and end) of the mangrove portions along the estuarine banks were obtained during a field expedition with the help of a Garmim EXTREX GPS, and the total mangrove area was confirmed through the use of satellite images.

The methods used for mapping the habitat sensitivity to oil followed the federal recommendations (MMA, 2004), which themselves were based on protocols from the NOAA (2002). The digitalization of information on mangrove areas and the compilation of data from cartographic databases were performed with the aid of ArcGIS 10 software.

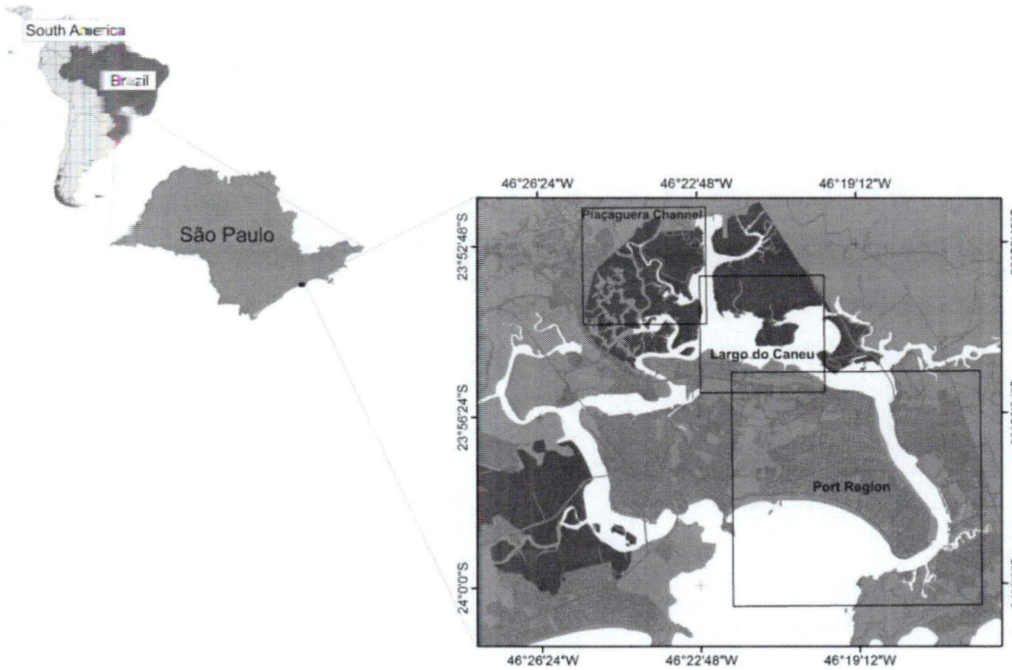

Figure 6. Location of study area – Santos and São Vicente Estuarine System. The black areas in detail corresponds to mangroves.

The organisms were represented by plots or polygons according to the information found in the literature. Each organism group received an icon and a reference number under the icon that was linked to a table containing the information on the species. The analysis of information obtained in the literature showed that the region closest to the Piaçaguera Channel presents the largest number of taxa, followed by the areas surrounding Largo do Caneu. These two areas present well developed mangroves that contain proper areas for feeding, nursing and breeding for several species, and which thus favor the presence of larger numbers of species in the area. In the study, the area around the port had the lowest number of taxonomic groups, which is due to the absence of mangrove areas and intense anthropic activities

As is shown in Table 6, this analysis was performed using the number of taxa (species or other taxonomic groups) found in each region. The Piaçaguera Channel was found to contain 261 taxa. The "Largo do Canéu" region is home to 252 taxa, and the Port region contains 230 taxa. Ecological indices were not used in this study due to the variations in the sampling methods used by authors (Krebs, 1998).

The number of endangered species may contribute to the area's increased environmental sensitivity to oil. At the Piaçaguera Channel, 34 endangered species were found, while at the other two sites (Largo do Caneu and Port Region), 31 and 22 endangered species were found, respectively.

Group		Number of species or taxonomic groups		
		Piaçaguera	Caneu	Porto
Birds	Wading Bird	3	8	0
	Waterfowl	6	5	0
	Raptors	10	5	0
	Passerine Birds	26	10	2
	Non-Passerine Birds	16	4	0
	Shorebirds	5	2	0
	Diving Birds	1	1	0
	Gulls/Terns	1	9	1
Submerged Aquatic Plants		6	6	0
Repitiles / Amphibians	Crocodiles	1	0	0
	Amphibians	5	0	0
	Snakes	2	4	0
Mammals	Terrestrial Mammals	6	1	0
	Sea otters	0	1	0
	Others	1	0	0
Marine Aquatic Invertebrates	Bivalves	0	1	0
	Gastropods	0	1	2
	Shrimp	0	6	6
	Crabs	41	41	41
	Other groups	0	15	21
Plankton		7	8	33
Fishes	Pelagic	105	105	105
	Demersal	19	19	19

When the species are displayed in the maps, it is evident that many species are more widely distributed along the mangrove areas, especially those of the Piaçaguera Channel (Figure 7). Thus, because Piaçaguera Channel appears to present more targets (taxonomic groups) that can be affected by oil in the case of a spill, this area should be considered highly sensitive. On the other hand, the Port region tended to be less sensitive to oil (Figure 8), whereas the Largo do Caneu presented an intermediate sensitivity level (Figure 9) compared to the Piaçaguera Channel.

Main Taxonomic Groups Found in the Different Sections of the Santos – São Vicente Estuarine System.

The results showed that, although mangrove areas are classified under the maximum ESI (ESI 10), some mangrove areas within a broad estuarine area may be more fragile than others, depending on the species inhabiting the region. Thus, the establishment of priority areas should take into account species richness and taxon distribution as well, and should not rely on the ESI alone.

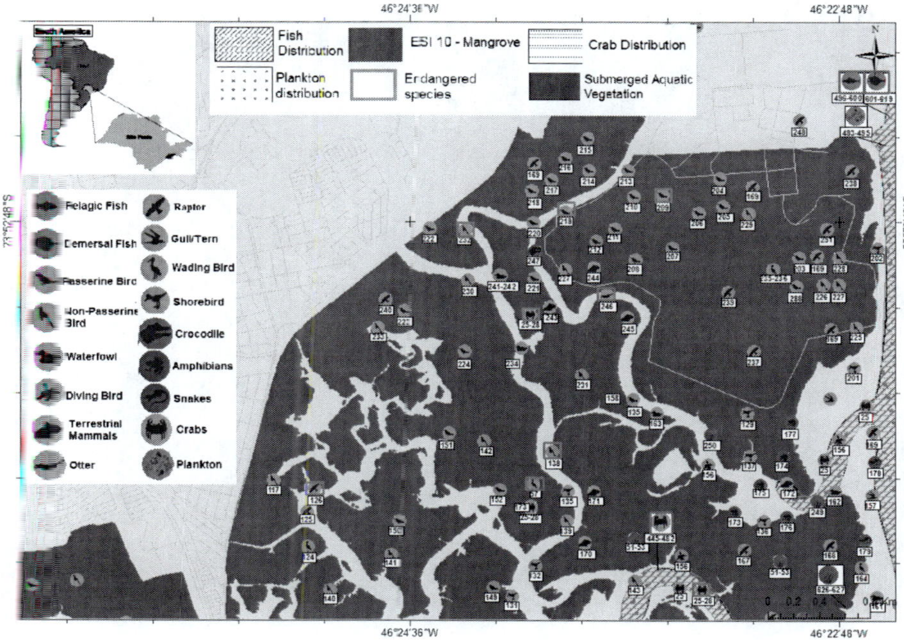

Figure 7. Detail of the ESI Map for the Piaçaguera Channel presenting the distribution of the main taxonomic groups in study area.

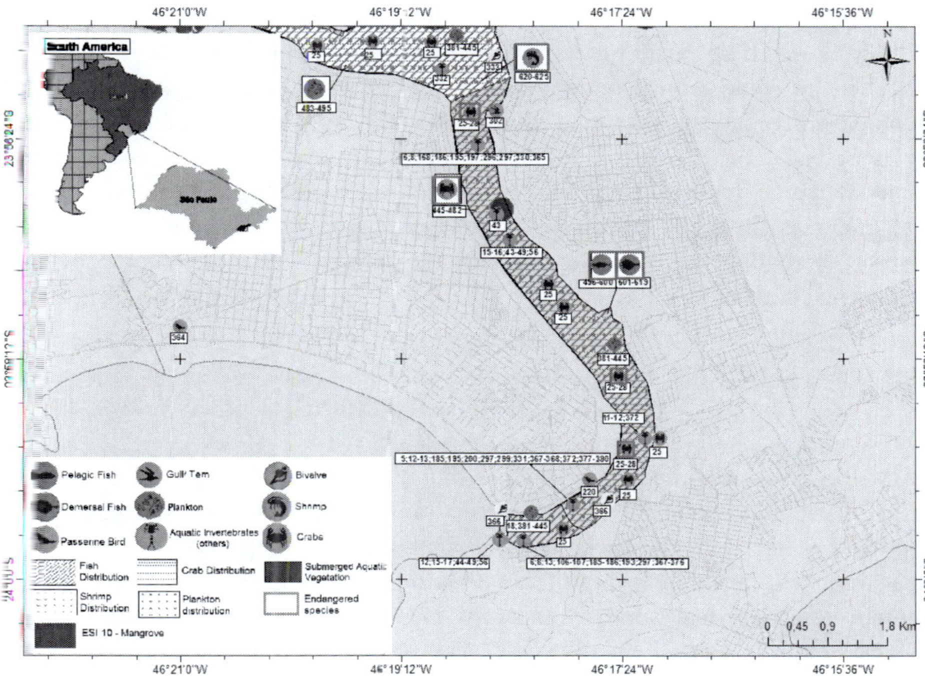

Figure 8. Detail of the ESI Map for the Port Region presenting the distribution of the main taxonomic groups in study area.

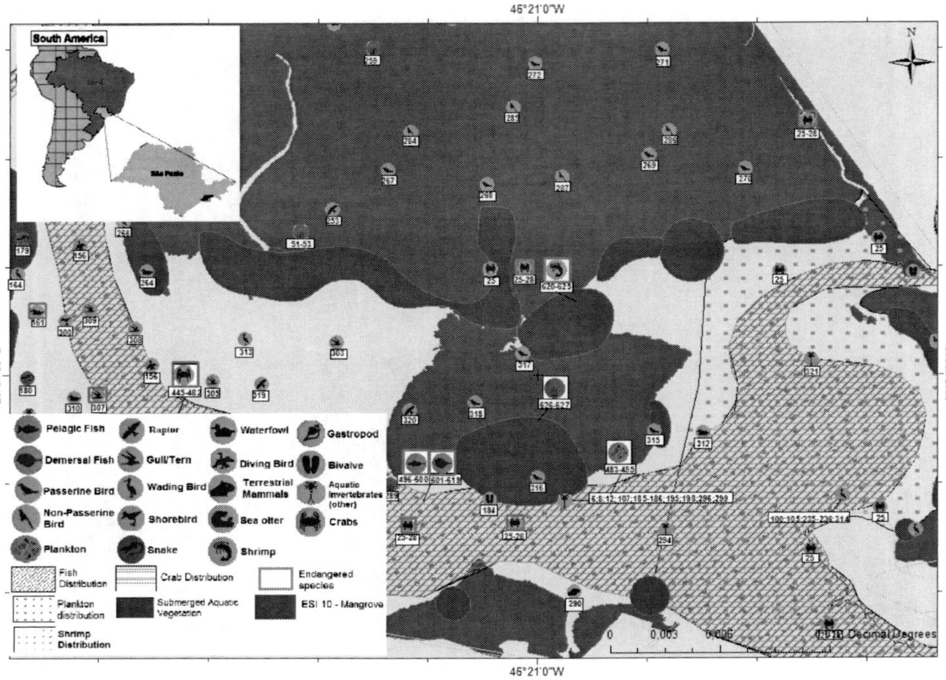

Figure 9. Detail of the ESI Map for Largo do Caneu presenting the distribution of the main taxonomic groups in study area.

The map showed that, even in areas classified under the same ESI, there could be a need to give priority to some critical regions due to the presence of biological resources (i.e., higher biological diversity, richness and/or density). Therefore, the way the biological data is currently represented on ESI Maps does not support an appropriate response plan for oil spills, particularly in terms the species or ecosystem protection. Thus, information on biological resources should be more appropriately displayed on ESI Maps or MEVOs as an important part of contingency plans. In doing so, they will provide more reliable information for environmental planners.

FINAL CONSIDERATIONS

Contingency plans are important strategies in the preparation of response actions for oil accidents. However, they should evolve by incorporating new approaches and technologies in order to provide the best possible tools for avoiding the negative impacts of oil.

In this sense, the new approaches presented in this chapter allow for a better emergency planning move the classical approach forward in an attempt to better mitigate possible environmental impacts caused by oil spills. The development of methods that integrate information on sensitivity and vulnerability of environments and organisms using tools such as GIS and numerical modeling can contribute to improved oil spill response planning and thus to better preservation of coastal environments. Moreover, these approaches are flexible and not exclusionary, and they allow for the incorporation of new tools. In light of current statistics, in which global oil production and international trade using shipping continue to increase, further research should be done to protect coastal environments and their biological resources. Similarly, more attention must be given to the socio-economic effects of oil spills. The development of new approaches and techniques aimed at protecting the environment from oil should be permanent and should involve human resources from different areas of knowledge.

REFERENCES

Afonso, C.M, A paisagem na Baixada Santista: urbanização, transformação e conservação, EDUSP, São Paulo, 2001.

Allan, S. E., Smith, B. W., Anderson, K. A., 2012. Impact of the Deepwater Horizon oil spill on bioavailable polycyclic aromatic hydrocarbons in Gulf of Mexico coastal waters. *Environ. Sci. Technol.*, 46, 2033-2039.

Cantagallo, C.; Milanelli, J. C. C.; Dias-Brito, D., 2007. Limpeza de ambientes costeiros brasileiros contaminados por petróleo: uma revisão. *Pan-American Journal of Aquatic Sciences*, 2(1): 1-12.

Castanedo, S., Juanes, J. A., Medina, R., Puente, A., Fernandez, F., Olabarrieta, M., Pombo, C., 2009. Oil spill vulnerability assessment integrating physical, biological and socio-economical aspects: Application to the Cantabrian coast (Bay of Biscay, Spain).

Concawe, 1998. Heavy fuel oils. Brussel. Available at: < http://www.accede.org/prestige/documentos/Tox_fuel_pesado.pdf>. Accessed on: 06 de fev. de 2009.

Diegues, A. C. 1996. Ecologia Humana e Planejamento em Áreas Costeiras. NUPAUB-USP, São Paulo, 192p.

Geyer, R. A., 1980. Marine environmental pollution. Elsevier Scientific Publishing Company, Amsterdam, Oxford, New York.

Gundlach, E.R. and M. O. Hayes, Vulnerability of Coastal Environments to Oil Spill Impacts, *Marine Technology Society Journal*, Vol. 12, pp. 18-27, 1978.

Harari, J. C.A.S. França and R. Camargo, Climatology and Hydrography of Santos Estuary. In: Perspective on Integrated Coastal Zone Managemente in South America. IST PRESS. pp. 147 – 160. 2008.

HÖFER, T. Tankships in the marine environment. *Environmental Sci. & Pollution Res.*, 5 (2), p. 97-104, 1998.

HUIJER. K. Trends in oil spills from tanker ships, 1995–2004. London, 2005.

Ihaksi. T., Kokkonen,T., Helle, I., Jolma, A., Lecklin, T., Kuikka, S., 2011. Combining Conservation Value, Vulnerability, and Effectiveness of Mitigation Actions in Spatial Conservation Decisions: An Application to Coastal Oil Spill Combating. *Environ. Manag.*, 47:802–813.

IPIECA. A guide to contingency planning for oil spills on water. IPIECA Report Series, Volume Two. 2nd edition. International Petroleum Industry Environmental Conservation Association. London, UK. 2000.

Junoy, J., Castellanos, C., Viétez, J. M., De La Huz, M. R., Lastra, M., 2005 The macroinfauna of the Galician Sandy beaches (NW Spain) affected by the Prestige oil-spil . *Mar. Pollut. Bull.*, 50, 526-536.

Kiesling, R. W., Alexander S. K., Web J. W. 1988. Evaluation of Alternative Oil Spill Cleanup Techniques in a *Spartina alterniflora* Salt Marsh. *Environmental Pollution*, 55: 221-238.

Kostka, J. E., Prakash, O., Overholt, W. A., Green, S. J., Freyer, G., Canion, A., Delgardio, J., Norton, N., Hazen, T. C., Huettel, M., 2011. Hydrocarbon-degrading bacteria and the bacterial community response in Gulf of Mexico Beach Sands Impacted by the Deepwater Horizon Oil Spill. *Appl. Environ. Microb.*, Nov., 7962-7974.

Krebs, C. J. Ecological methodology. Addison-Wesley educational Publishers, 1998.

Linden, O., Elmgren, R., Boehm, P., 1979. The *Tsesis* oil spill: its impact on the coastal ecosystem of the Baltic Sea. Ambio, vol.8, n.6, 244-253.

Lopes, C. F., Milanelli, J. C. C., Poffo, I. R. F., 2007. Ambientes costeiros contaminados por óleo: procedimentos de limpeza – Manual de orientação. Secretaria de Estado do Meio Ambiente, São Paulo. Available at: <http://www.cetesb.sp.gov. br/userfiles/file /emergencias-quimicas/acidentes/ambientes-costeiros.pdf>. Accessed on March 16, 2013.

Mendonza-Cantú, A., Heydrich, S. C., Cervantes, I. S., Orozco, O. O., 2011. Identification of environmentally vulnerable áreas with priority for prevention and management of pipeline crude oil spills. *J. of Environmental Management,* 92, 1706-1713.

Michel, J., Hayes, M. O., Brown, P. J. Application of an Oil Spill Vulnerability Index to the Shoreline of Lower Cook Inlet, Alaska. *Environmental Geology*, Vol. 2, N°.2, pp. 107-117. 1978.

MMA, Ministério do Meio Ambiente. Especificações e Normas Técnicas para a Elaboração de Cartas de Sensibilidade Ambiental para derramamentos de óleo, Brasília, 107p 2004. Available at: <http://www.mma.gov.br/seguranca-quimica/cartas-de-sensibilidade-ao-oleo/metodologia>. Accessed on March 17, 2013

Muler, M., Romero, A.F., Riedel, P. S., Perinotto, R. R. C., 2011. Ações de resposta para emergência em caso de derrames de óleo no mar e proposta de implementação de sistema de informação voltado à sensibilidade ambiental para o litoral sul paulista, Brasil. *Revista da Gestão Costeira Integrada,* 11 (4), 397-407.

NOAA (National Oceanic Atmospheric Administration), Enviromental Sensitivity Index Guidelines, Version 3.0. NOAA Technical Memorandum NOS ORCA 115, Seattle: Hazardous Materials Response and Assessment Division, *National Oceanic and Atmospheric Administration.* 89 p. + appendices, 2002.

Pezeshki, S. R., Hester b, M. W., Lin, Q., Nyman, J. A. 2000. The effects of oil spill and clean-up on dominant US Gulf coast marsh macrophytes: a review. *Environmental Pollution*, 108: 129-139.

POFFO, I.R.F.; XAVIER, J.C.M.; SERPA, R.R. A História dos 27 anos de vazamento de óleo no litoral norte do Estado de São Paulo (1974-2000*). Revista Meio Ambiente Industrial*, n° 30, p.98-104, 2001.

Ramseur, J.L., 2010, Deepwater Horizon Oil Spill: The Fate of the Oil. Congressional Research Service, <http://www.fas.org/sgp/crs/misc/ R41531.pdf>, Accessed on 20-02-2014.

Romero, A. F., Riedel, P. S., Milanelli, J. C. C., Lammardo, A. C. R., 2011. Mapa de Vulnerabilidade Ambiental ao Óleo – Um Estudo de Caso na Bacia de Santos, Brasil. *Revista Brasileira de Cartografia*, 63/03.

Romero, A. F., Abessa, D.M.S., Fontes, R.F.C., Silva, G.H. Integrated Assessment for Establishing an Oil Environmental Vulnerability Map: Case Study for the Santos Basin Region, Brazil. *Marine Pollution Bulletin.* 74 (2013) 156–164.

Romero, A.F., Abessa, D.M.S. Mapa de Procedimentos de Limpeza e Ações de Combate a Derramamentos de Óleo. Brazilian *Journal Of Aquatic Science And Technology*, 2014 (in press).

Sant'anna-Netto. J. L, Ritmo Climático e as Chuvas na Zona Costeira Paulista, Dissetação de Mestrado, Faculdade de Filosofia, Letras e Ciências Humanas da Universidade de São Paulo, 1990.

Santos, E. O. A Baixada Santista: Aspectos Geográficos. Edusp, Vol. 1, chap.4, pp. 95-150, 1965.

Silva, G. H., Lima, S. O. F. L., Araújo, S. O., Gomes, C. C., 2012. Mapeamento da Vulnerabilidade Ambiental a Derrames de Óleo em Ambientes Costeiros. Anais do I Congresso Brasileiro de Avaliação de Impacto. São Paulo. Available at: <http://avaliacaodeimpacto.org.br/wp-content/uploads/2012/10/085_impact_petroleo.pdf>. Accessed on March 18 2013.

Tessler, M. G.; Goya, S. C., 2005. Processos Costeiros Condicionantes do Litoral Brasileiro. *Rev. Depart. Geogr.*, 17, 11-23.

Thomas, M. L. H., 1973. Effects of Bunker C oil on intertidal and lagoonal biota in Chedabucto Bay, Nova Scotia. *J. Fish. Res. Board. Can.*, 30, 83-90.

Thomas, M.L.H., 1978. Comparison of oiled and unoiled intertidal communities in Chedabucto Bay, Nova Scotia. *J. Fish. Res Board. Can.,* 35, 707-716.

UNCTAD. Review of maritime transport 2011. New York, Geneva. UNCTAD/RMT/2011 - UNITED NATIONS PUBLICATION - Sales no.E.11.II.D.4, 2011.

USEPA. United States Environmental Protection Agency. Office of Emergency and Remedial Response. Understanding Oil Spills And Oil Spill Response: Understanding Oil Spills in Freshwater Environments. 1999. Available in http://www.epa.gov/ oem/content/learning /pdfbook.htm

Van De Wiel, G.and J.R Van Dorp, An Oil Outflow Model for Tanker Collisions and Groundings. *Ann. Oper. Res.* Vol. 187, pp 279–304, 2011.

Ventikos, N. P., Vergetis, E., Psaraftis, H. N., Triantafyllou, G., 2004. A high-level synthesis of oil spill response equipment and countermeasure. *Journal of Hazardous Materials* 107 51–58.

In: Oil Spills
Editor: Adam Clifton
ISBN: 978-1-63321-515-3

Chapter 6

USE OF STABLE AND RADIOGENIC ISOTOPES IN OIL SPILL CASES

V. G. Barros[1*], T. M. N. Oliveira[1], R. Spandre[2], G. M. Zuppi[3] and J. Rapaglia[4]

[1]University of UNIVILLE, Dept o Eng Ambiental, Campus Universitário Bom Retiro Joinville, Santa Catarina, Brazil
[2]Dipartimento di Scienze dell 'Uomo e dell' Ambiente, Università di Pisa, Italy
[3]Dipartimento di Scienze Ambientali, Università Ca' Foscari, Dorsoduro, Venezia, Italy. In Memorian
[4]Department of Biology, Sacred Heart University, Fairfield, CT, US

ABSTRACT

Environmental tracers, as isotopes, have been used to reveal origin, movement and dispersion of compounds because they behave the same as environmental processes. These isotopes are useful tools for understanding dynamics of natural systems, providing a unique tool to trace the biogeochemical cycles within ecosystems. The hydrological complex of Babitonga Bay (Brazil) forms a vast environmental system; hosting the last great expanse of mangrove forests in the southern hemisphere. Mangroves are among the most productive ecosystems on earth. Effects of an oil spill on mangrove ecosystems were studied in Babitonga Bay using lead and carbon isotopes. Samples of the spilled oil, as well as sediment and water samples, were obtained nine months after the accident, at the time of the salvage operation. Isotopic composition of the oil was utilized to trace the extent of the environmental pollution. In addition, isotopes of lead and carbon allowed for the identification of areas where oil was present. Contaminated sediments exhibited an isotopic composition ($^{206}Pb/^{207}Pb$ and $^{208}Pb/^{206}Pb$) close to that of the oil spilled. $\delta^{13}C$ data confirmed these results. The results of this investigation suggest that lead isotope ratios can be very useful in the field of environmental forensics.

* Corresponding author: Email: vgbarros@gmail.com.

1. Introduction

Brazil is experiencing fast economical growth. This growth includes an increase in both production and transportation of raw materials and finished products. Brazilian harbors are located in estuaries, which are highly susceptible to changes in the ecosystem caused by increased economic production. Estuaries and bays are very important ecosystems both as a source of food production, and for recreational, aesthetic and cultural purposes (Barros et al., 2010). Despite their importance, these areas are heavily impacted by contaminants from many different sources. Estuaries and bays constitute natural reaction zones in which heterogeneous processes may affect the ecological equilibrium due to local changes in the biogeochemical conditions. In addition to natural sources, anthropogenic inputs provide significant contributions to these areas. In Babitonga Bay (BB), Brazil, ,many factors lead to sediment pollution (CREMER et al., 2006); the major players being: (i) activities at São Francisco do Sul Harbour, (ii) inadequately treated domestic sewage, (iii) industrial activity at Joinville, and (iv) other contamination transport from tributaries (Barros, Grace et al., 2008, Barros et al., 2010, Vaz et al., 2010). Estuarine systems are characterized by large amounts of organic matter, therefore the composition of stable carbon and nitrogen isotopes is useful in estuarine studies as a technique for understanding processes (Barros et al., 2010, Dehairs et al., 2000, Carreira et al., 2002, Graham et al., 2001, Goñi et al., 2003). The composition of stable carbon isotope is a useful tracer in estuarine studies as terrestrially derived material is isotopically lighter than marine materials when not under the influence of C4 plants. Terrestrial material has $\delta^{13}C$ values between −30 and −24‰, while material of marine origin has values between −22 and −18‰.

The interpretation of isotope ratios can be complicated by the fact that material, especially bulk material, often represents mixtures of carbon from several different sources, and thus the isotope signatures reflect integration from these different sources. Moreover, systems have variable inputs in space and time and, thus, it is difficult to separate the signal of multiple end-members from the effects of matter alteration and transport (Kendall and McDonnell, 1998; Dittmar et al., 2001; Gonneea et al., 2004).

In order to distinguish between anthropogenic and natural inputs, radiogenic isotopes of lead represent a potential indicator. Lead (Pb) is a natural constituent of the Earth's crust, which is commonly found in soils, plants, and water at trace levels. However, occurrence of metallic lead in nature is rare. It is a toxic heavy metal that can enter the human body through inhalation and ingestion of contaminated air and water, soil, and food (Cheng and Hu, 2010). It is also fairly immobile in the environment. This study was conducted within an environmental investigation. Especially in a forensic environmental investigation, it is important to determine the source of pollution and/or to characterize the transport history of the pollutants. Recognizing the sources of lead pollution can be difficult because of the large number of potential sources in typical urban settings. These include emissions from industries, exhaust from e.g., fuel combustion, sewage, atmospheric deposition, and even natural geological materials (bedrocks and soils). While the sources of Pb and the contribution from each source cannot be inferred from the total Pb concentrations measured in environmental media (e.g., air, water, and soil), it is possible to distinguish between natural and anthropogenic sources using Pb isotope ratios.

A case in which the use of Pb isotopes may present clues to human impac on the environment involves a shipping accident which occurred in the summer of 2008 leading to a large oil spill, greatly impacting the BB ecosystem. A salvage operation was conducted eight months after the spill to recover the shipwreck. This operation caused further loss of oil yet provided an opportunity to collect samples of oil remaining in the holds of the ship. One month after the salvage operation, a sampling campaign was performed in order to search for residues of this spill. The major objectives of the present study, therefore, are to characterize the different isotopic compositions of carbon, nitrogen and lead in BB, and to elucidate on the origin of Pb, N and C in the sediments. Lead, nitrogen and carbon isotopic compositions were determined for bulk sediments and residues in order to identify the different (natural or anthropogenic) inputs of oil. Many studies have shown correlations between lead and sediments in estuarine areas, but using different approaches from those presented here. In the current case, lead was chosen as an additional tracer to identify the origin of the oil, as the isotopic composition of Pb is not significantly affected by physico-chemical fractionation processes (Komárek et al., 2008).

2. TRACERS

Environmental tracers have been used to reveal origin, movement and dispersion of other compounds because they behave the same as environmental processes. Environmental isotopes are the naturally occurring isotopes of elements found in abundance in the environment. Both stable and radioactive, these isotopes are useful tools for understanding dynamics of natural systems, providing a unique tool to trace the biogeochemical cycles within ecosystems. Isotopes provide information about the natural cycles of elements and compounds in hydrologic systems. Indeed isotopic tracers contribute to the understanding of water and element origins, recharge rates, mean residence time, (geo)physical and (geo)chemical processes. In addition they can determine important physical parameters such as water mixing, evaporation, water-rock interaction and further provide additional support to the geochemical and physical hydrogeological data. Measured tracer-response produced from diffuse emission of an unknown quantity of tracer provides an efficient method of obtaining necessary information.

2.1. Isotopes

Isotopes are elements that have the same atomic number, Z, but different number of neutrons, N, therefore having a different mass number A, ($A=Z+N$). There are two kinds of isotopes, stable and radioactive, both of which can be used as tracers. Multiple isotopes of elements may exist naturally in the environment. The ratio of different iostopes of an element often give important information as to the source of the water particle. Indeed source isotopic ratios are often very constant, and there can be great variation in the ratios between different sources. The variability that occurs is the function different rates of reaction in thermodynamic reactions for different molecular species. Therefore, isotopic ratios can be a very effective tool in understanding environmental processes.

3. Notation

When identifying an isotope of an element a superscript number to the left of the element designates the sum of protons plus neutrons in the isotope. So, ^{18}O defines one oxygen which has a Z= 8 and an N= 10 giving a mass number A=18. Others examples are:

Isotope	Z	*N*	A
^{16}O	8	8	16
^{18}O	8	10	18
^{2}H or D	1	1	2
^{3}H	1	2	3

3.1. Terminology

(R) - Isotope ratio or abundance
Is defined by the ratio:

$$R = \frac{Abundance\ of\ rare\ (or\ heavier)\ isotope}{Abundance\ of\ abundant\ (or\ lighter)\ isotope} \quad (1)$$

3.2. Standards

The isotopic ratios of materials are analyzed on mass spectrometers and are usually reported relative to international reference standards. Some isotope standards are listed in table 1.

The δ values of each of the standards conventionally have been defined as 0‰. In other words, the standards will provide a baseline value and the samples will then be reported as either lighter or heavier than the standard values (±X‰, see below for further discussion) $\delta^{2}H$ and $\delta^{18}O$ values are usually reported relative to the SMOW standard (Standard Mean Ocean Water; Craig, 1961) or the equivalent VSMOW (Vienna-SMOW) standard. $\delta^{13}C$ and $\delta^{18}O$ values of low-temperature carbonates values are reported relative to either the PDB (Pee Dee Belemnite) or the equivalent VPDB (Vienna-PDB) standard. Use of VSMOW and VPDB is supposed to imply that the measurements were calibrated according to IAEA (International Atomic Energy Agency) guidelines for expression of δ values relative to available reference materials on normalized per mil scales (Gonfiantini, 1978; Gonfiantini et al., 1995; Coplen, 1994; 1995; 1996). This certifies that VSMOW and VPDB are virtually identical to the SMOW and PDB standards.

3.3. δ - Delta Notation

It is important to observe that absolute ratios are, in most cases, not as significant as the changes in ratios occurring during chemical and physical reactions. For this reason, the composition or abundance is usually reported as a deviation of the isotope ratio of a sample (A) relative to that of a reference sample or standard, (st). This is translated and reported by delta "δ", in parts per thousand (‰), which is a relative value that indicates the amount of the rare or heavy isotope by the most abundant or light species, or:

$$\delta = \left(\frac{R_A}{R_{st}} - 1\right) \cdot 10^3 \quad (‰) \qquad (2)$$

where:

R_A is the ratio of the heavy to light isotope (e.g., $^{18}O/^{16}O$; $^{13}C/^{12}C$) in the sample, and, R_{st} is the ratio of the heavy to light isotope in the standard.

"δ" values can be negative or positive, and according to Mook (IAEA 2001), a negative value indicates lower abundance of the rare isotope in these samples compared to the standard while a positive value points to higher abundance.

Table 1. Abundance ratios and reference standards for some environmental isotopes

Isotope	Ratio measured	Reference Standard	Abundance (%)
^{1}H	$^{2}H/^{1}H$	VSMOW	99.985
^{2}H			0.0155
^{12}C	$^{13}C/^{12}C$	VPDB	98.892
^{13}C			1.108
^{14}N	$^{15}N/^{14}N$	Atmospheric N_2	99.635
^{15}N			0.365
^{16}O	$^{18}O/^{16}O$	VSMOW, or VPDB	99.759
^{18}O			0.204

3.4. Isotope Fractionation

The chemical and physical properties of the isotopes of a certain element are slightly different in function as well as different in nuclear mass. The consequences of these different properties are defined by Mook (2001), as:

1. The heavier isotopic molecules have lower mobility. The kinetic energy of a molecule is solely determined by temperature: $kT = ½mv^2$ (k = Boltzmann constant, T = absolute temperature, m= molecular mass, v = average molecular velocity). Therefore, the value $½mv^2$ must remain the same regardless of their isotope content. Hence, in order to keep the value the same, molecules with larger molecular mass necessarily have a smaller average molecular velocity. Some practical consequences of this quality are:

a. heavier molecules have a lower diffusion velocity;
b. the collision frequency with other molecules - the primary condition for chemical reaction - is smaller for heavier molecules; this is one of the reasons why, as a rule, lighter molecules react faster.
c. The heavier molecules generally have higher binding energies. For example, there is about a 2 kcal/mole difference in ZPE (zero point energy) associated with the breaking of the H-H bond compared to the D-D bond, figure 1.

During thermodynamic processes the different isotopes will react with different rates. As stated by Clark and Fritz (1997), the result is a disproportionate concentration of one isotope compared with the other isotope on one side of the reaction. It is expressed by the fractionation factor "a" which is the ratio of the isotopic ratios for both the reactant and product:

$$\alpha_A(B) = \alpha_{B/A} = \frac{R_B}{R_A} \qquad (3)$$

Instead of using α values, which are small, a deviation of α from 1 is used instead of the fractionation factor. This deviation, termed fractionation, is defined by:

$$\varepsilon_{B/A} = \alpha_{B/A} - 1 = \frac{R_A}{R_B} - 1 \text{ (x } 10^3 \text{ ‰)} \qquad (4)$$

ε represents the enrichment ($\varepsilon > 0$) or the depletion ($\varepsilon < 0$) of the rare isotope in B with respect to A. As described by Mook (2001), in the one-way process ($A \rightarrow B$), ε is the change in isotopic composition, in other words: the new isotopic composition compared to the old.

There are two different types of isotope fractionation, which are a function of the processes that cause them. If the fractionation is a result of an irreversible (one-way) physical or chemical process, (in a non-equilibrium system) kinetic fractionation is taking place. On the other hand, if the fractionation is a result of an (thermodynamic) equilibrium reaction, the second type of fractionation, equilibrium (or thermodynamic) fractionation, is proceeding. Regarding the degree of these two types of fractionation, Kendall and McDonnell (1998) state that the magnitude of a kinetic isotope fractionation depends on the reaction pathway, the reaction rate, and the relative bond energies of the bonds being severed or formed by the reaction. Kinetic fractionations, especially unidirectional ones, are usually larger than the equilibrium fractionation factor for the same reaction in most low-temperature environments. As a rule, bonds between the lighter isotopes are broken more easily than equivalent bonds of heavier isotopes. Hence, the light isotopes react faster and become concentrated in the products, causing the residual reactants to become enriched in the heavy isotopes. In contrast, reversible equilibrium reactions can produce products heavier or lighter than the original reactants.

3.5. Radiogenic Isotopes

It is known that radiogenic isotopes do not have the same mass as the parent nucleus. By conservation of mass and charge, the alpha decay equation can be written:

$$ {}^{A}_{Z}A \rightarrow {}^{A-}_{Z-2}B + {}^{4}_{2}He $$

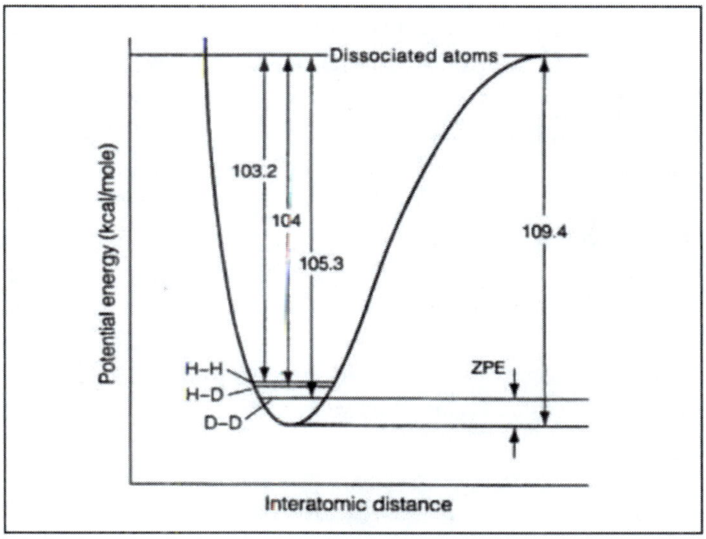

Figure 1. Inter atomic distance versus potential energy for stable hydrogen isotopes of a molecule. Higher ZPE (zero point energies) result in molecules being less stable. Modified from Kendall McDonnell (1998).

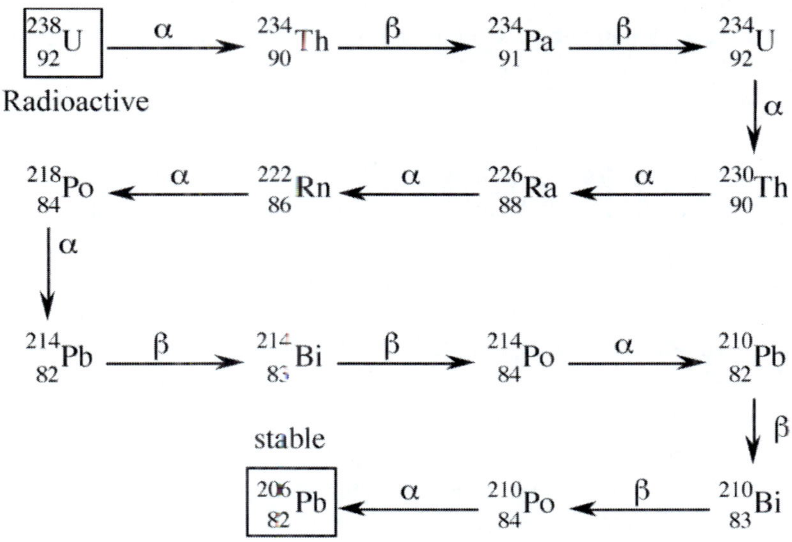

Figure 2. Radioactive chain to 238U identifying the types of radioactive decay.

Lead has four stable, naturally occurring isotopes (natural % abundance): ^{204}Pb (1.4%), ^{206}Pb (24.1%), ^{207}Pb (22.1%) and ^{208}Pb (52.4%). Three of the isotopes of lead are are the end products of complex decay of radioactive chains: ^{206}Pb by ^{238}U, ^{207}Pb by ^{235}U, and ^{208}Pb by ^{232}Th, also known as the Uranium-Thorium decay series (Rutgers van der Loeff, 2001). The corresponding half-lives of these decay schemes vary markedly: 4.47 x 10^9 for ^{238}U, 7.04 x 10^8 for ^{235}U and 1.4 x 10^{10} years for ^{232}Th. Assuming the chains are in secular equilibrium (given the length of geological time), U and Th decay directly to Pb isotopes. Each is reported relative to ^{204}Pb, the only non-radiogenic stable isotope. The advantage with lead isotopes is that the results are naturally correlated since the two decay schemes are chemically similar and it is only the decay constant that differs. A radioactive chain involves decay with α and β radioactivity. α radioactive releases He nuclei. Figure 2 shows the path of radioactive decay to 238U chain presenting loss of α and β particle.

The ranges of isotopic ratios for most usual materials are 14.0-30.0 for $^{206}Pb/^{204}Pb$, 15.0-17.0 for $^{207}Pb/^{204}Pb$ and 35.0-50.0 for $^{208}Pb/^{204}Pb$, although numerous examples outside these ranges are described. According to USGS (2004), because U, Th and Pb have different geochemical behaviors, the fact that the Pb isotopic composition of any material is the composite of the three independent decay chains creates the potential for greater variability of isotopic values in minerals of a single rock relative to that for the Rb-Sr system. Also, atmospheric Pb tends to become concentrated in the uppermost, organic-rich soil horizons, while Pb in ground water should be dominated by that derived from rock weathering.

4. Site Description

The hydrological complex of Babitonga Bay (BB) is located 45 km northeast from the city of Joinville, which is the principal (capital) city of Santa Catarina State (Southern Brazil), Figure 3. The drainage basin of the bay consists of a vast environmental complex, where agriculture and shellfish farming as well as a variety of industries coexist with a unique natural area of Brazilian rain forest. Babitonga Bay, which houses the largest expanse of mangrove forests left in the southern hemisphere (160 km^2), is connected to the Atlantic Ocean via an inlet with a width of 1850 m and is characterized as a salt-wedge estuary. Tides are mixed, semidiurnal with a mean range of 1,80 m (Schettini and Carvalho, 1996). This oscillation is very important for the flushing of the entire system. The region is characterized by a humid subtropical climate with an average rainfall of 2000 mm/year. There are two seasons: summer (from November to April) and winter (from May to October). During summer the weather is characterized by high temperature and humidity (annual meam: 20,5ºC and 87,18%), with intense precipitation (annual mean 2000mm) caused by local convection processes strictly controlled by the regional orography. During winter the influence of polar air-masses leads to a decrease in both temperature (2,6ºC) and precipitation.

The complex encompasses two main basins: Cubatão and Cachoeira. The largest stream discharging into the BB is the Cubatão River, which has a drainage basin of 483.8 km^2 and a length of more than 75 km. Its source is located in the Serra do Mar at an altitude of 1300m. The mean discharge at the mouth of the river is 17.7 m^3/s but is marked by rapid fluctuations due to intense precipitation events on the steep slopes of the Serra do Mar. In addition to being used as a public water supply, Cubatão waters are also withdrawn for industrial and agricultural purposes (IBAMA, 1998). The Cachoeira River basin (84.82 km^2) is located entirely within the urbanized area of Joinville, and flows into the Saguaçu Lagoon. Its hydraulic head is located at an altitude of 40 m, with a mean riverbed elevation of about 10 m. The mouth is characterized by estuarine circulation and is home to residual mangrove forests. There are little data concerning the Cachoeira River discharge; however it is believed to range from 3 to 5 m^3/s. Water quality is very poor, mainly due to the input of mostly untreated domestic and industrial effluents. Heavy loading of untreated wastes leads to very low concentrations of dissolved oxygen observed in this river. Two other relatively pristine rivers discharge into BB. The Palmital River basin drains an area of 357.60 km^2 and the Parati River drains an area of 72.20 km^2: their estuaries receive contributions from several streams, all of which are influenced by tides and heavily forested with mangroves.

5. Geological Context

5.1. Santa Catarina Granulite Complex - SCGC

The high-grade terranes in Southern Brazil are most visible in the SCGC, with an exposed area of about 6,000 km^2 (Hartmann et al., 1979). The complex is unique; it was not internally impacted by thermal effects or by ductile deformations during the orogenesis of the Neoproterozoic Brasiliano cycle However, it was impacted by the intrusion of Brasiliano granites. The Santa Catarina Granulite Complex is the oldest tectonic unit in the Catarinense Shield (Guadagnin et al., 2010). This complex is composed of tonalitic, granodioritic, and trondhjemitic orthogneisses interspersed with mafic and ultramafic rocks. Metasedimentary rocks such as quartzites and banded iron formations are also present, albeit to a lesser extent. The first mantle–crust differentiation of the Santa Catarina Granulite Complex is of Archean age (2.8 Ga (TDM model age)). Meanwhile, tectonic stabilization occurred in the Paleoproterozoic (1.9 Ga (K-Ar in biotite)). The Santa Catarina Granulite Complex metamorphosed into a granulite facies during two events. The first event occurred at 2675±12 Ma, and the second event occurred at 2168±18Ma (U-Pb SHRIMP) when the magmatic protoliths crystallized at 2716±17Ma (Hartmann, et al. 2000). Hartman et al. 2000, interpreted the SCGC as a stable Craton after 1.9 Ga, and the complex remained an undeformed block during the orogenesis of the Brasiliano/Pan-African Cycle (Luis Alves Microplate).

6. Materials and Methods

Sampling was authorized by public prosecutor (PP) nine months after the oil spill. A preliminary campaign, also authorized by the PP, was performed one month before the main investigation. This initial campaign took place when the salvage operation was conducted, in order to collect three samples of the oil, which remained trapped for 8 months inside the hull of the capsized ship. The salvage operation led to further loss of oil inside the bay. Sample sites were chosen as a function of the oil slicks that were formed after the spill. Samples collected at these sites were analyzed by Centro de Pesquisas Geocronológicas - University of São Paulo – USP (Pb isotopes) and Centro de Energia Nuclear na Agricultura – University of

São Paulo – USP (C isotopes). BB bottom sediment samples were collected from the upper 15 cm of the benthic layer with a Petit Ponar sediment grab and sealed immediately. Beach sands from the upper 15 cm were collected with a plastic grab and sealed immediately. Within BB organic matter and mud are the dominant sediment fractions, while beaches were mainly composed with sand.

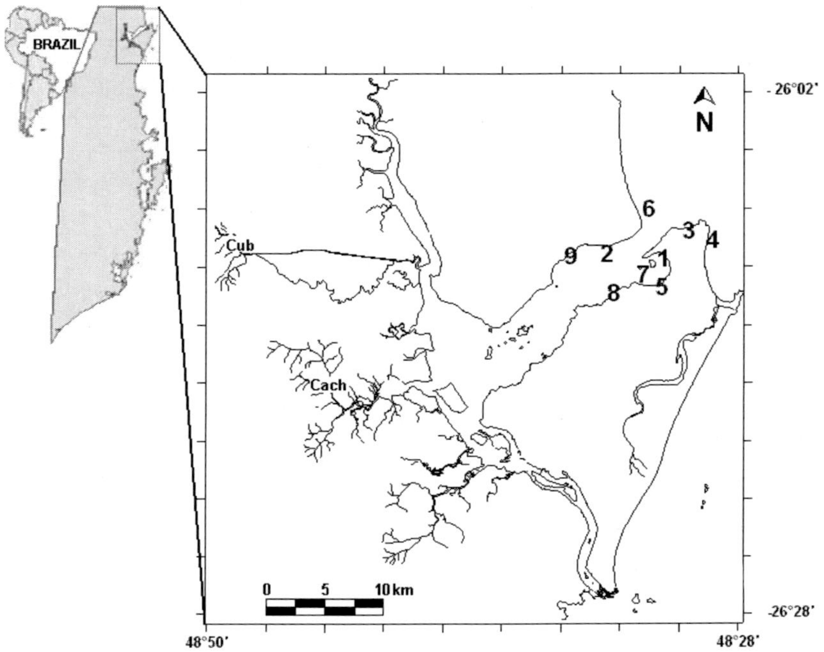

Figure 3. Babitonga Bay rispect to Brazil and sample points.

In order to complete the Pb isotopic analysis, each sample was washed ultrasonically in triple distilled water before chemical degradation. The sediment was completely dissolved using 3 ml of 6 N HCl plus 2ml of HNO_3 with a Parr® Teflon® bomb and then further dissolved in 2 ml of 0.7 N HBr. The sample was then passed through an ion exchange column filled with Dowex 1X8 AG anion exchange resin. After treatment within the column, the solutions were evaporated and the remaining residues were further dissolved in 0.7 N HBr and passed again through the same (Dowex 1X8 AG) ion exchange column as described above. Pb was separated using HBr+HCl chemistry in Dowex 1X8 AG ion exchange resin. Lead was loaded onto a Re ribbon, using silica gel plus H_3PO_4 and analyzed by a fully automated VG 354 Micromass multicollector thermal ionization mass spectrometer. Analyses using the "National Bureau of Standards" standard NBS 981 yielded mass discrimination and fractionation corrections of 1.0024 ($^{206}Pb/^{204}Pb$), 1.0038 ($^{207}Pb/^{204}Pb$) and 1.0051 ($^{208}Pb/^{204}Pb$); the combination of these uncertainties and within-run uncertainties are typically 0.15%–0.48% for $^{206}Pb/^{204}Pb$, 0.13%–1.07% for $^{207}Pb/^{204}Pb$ and 0.10%–0.45% for $^{208}Pb/^{204}Pb$, all at the 2σ (95%) confidence level. The total Pb blank contribution, <1 ng, is negligible. Classical Pb concentration in sediments and sands were performed using inductively coupled plasma - atomic emission spectrometry (ICP-AES), following USEPA 6010C.

Table 2. Isotopic compositions of BB sediments and oil samples. L is for leachate and R for residue. Numbers in samples S1, S2, etc. are sample points, Figure 3

Sample	$^{206}Pb/^{204}Pb$	Error % (1σ)	$^{207}Pb/^{204}Pb$	Error % (1σ)	$^{208}Pb/^{204}Pb$	Error % (1σ)	$^{206}Pb/^{207}Pb$	Error % (1σ)	$^{208}Pb/^{206}Pb$	Error % (1σ)	$\delta^{15}N$	$\delta^{13}C$
Oil1	18.654	0.056	15.613	0.058	38.507	0.058	1.195	0.009	2.065	0.008	8.24	-24.57
Oil2	18.347	0.225	15.595	0.223	38.152	0.226	1.177	0.021	2.080	0.013	8.59	-24.15
Oil+seaw	18.416	0.006	15.638	0.006	38.269	0.006	1.178	0.001	2.078	0.002	8.64	-26.31
S1 - L	18.345	0.007	15.688	0.007	39.040	0.007	1.169	0.002	2.128	0.002	nd	nd
S1 - Res	17.783	0.023	15.650	0.025	37.762	0.024	1.136	0.004	2.124	0.002		
S2 - L	17.981	0.008	15.684	0.009	38.918	0.011	1.147	0.002	2.165	0.004	4.22	-24.69
S2 - Res	17.475	0.006	15.544	0.006	37.835	0.006	1.124	0.001	2.165	0.002		
S3 - L	18.333	0.016	15.704	0.017	39.411	0.02	1.167	0.005	2.150	0.009	nd	nd
S3 - Res	22.893	0.004	16.284	0.005	42.810	0.005	1.406	0.001	1.870	0.002		
S4 - L	18.225	0.01	15.692	0.011	38.996	0.012	1.161	0.002	2.140	0.003	nd	nd
S4 - Res	17.468	0.006	15.685	0.005	37.673	0.006	1.114	0.001	2.157	0.001		
S5 - L	18.102	0.006	15.672	0.006	38.942	0.007	1.155	0.001	2.151	0.001	4.92	-21.71
S5 - Res	18.312	0.005	15.695	0.005	39.620	0.005	1.167	0.001	2.164	0.001		
S6 - L	18.566	0.016	15.702	0.015	39.370	0.017	1.182	0.003	2.121	0.002	nd	nd
S6 - Res	18.343	0.01	15.688	0.01	40.276	0.01	1.169	0.002	2.196	0.001		
S7 - L	18.132	0.054	15.696	0.05	39.023	0.053	1.155	0.006	2.152	0.007	10.57	-10.73
S7 - Res	17.310	0.056	15.662	0.056	37.414	0.057	1.105	0.006	2.162	0.005		
S8 - L	18.276	0.056	15.674	0.056	39.115	0.057	1.166	0.004	2.141	0.005	nd	nd
S8 - Res	18.113	0.025	15.686	0.024	37.954	0.024	1.155	0.004	2.096	0.003		
S9 - L	18.163	0.088	15.706	0.096	39.009	0.101	1.157	0.009	2.147	0.012	9.95	-2.07
S9 - Res	19.230	0.023	15.797	0.025	38.369	0.025	1.217	0.005	1.995	0.004		

Samples for carbon ($\delta^{13}C$) and nitrogen ($\delta^{15}N$) isotopic compositions in sediments and sands were initially washed with 0.1 N HCl to eliminate carbonates or shell fragments. Isotope ratios were determined in organic samples by the combustion of a known mass of the glass fiber filters in an Elemental Analyser (Carlo Erba EA 1400) in line with the mass spectrometer (Finnigan Mat Delta Plus). Results are expressed in ‰V-PDB and Atmospheric Nitrogen standards (Gonfiantini, 1978, 1995). Analytical errors are: 0.3‰ for $\delta^{13}C$ and 0.5‰ for $\delta^{15}N$. The analytical precision values were obtained based on duplicate measurements of internal calibrated standards used at the laboratory as well as samples from the present study (Gonfiantini et al., 1995).

7. Results and Discussion

The samples analyzed represent the surface layer of the sedimentary pile at the sites (Figure 3). Pb, N and C isotopic composition was determined in nine samples chosen by the public prosecutor, and are presented in table 2. Pb concentration is presented in table 3.

7.1. Lead

BB has a history of contamination by heavy metals (Cremer et al., 2006, Vaz et al., 2010, de Oliveira et al., 2010). Previous work found lead concentrations to be high in BB sediments (Cremer et al., 2006, Ahmed and Häder, 2010). Table 3 presents Pb concentration at the sites sampled.

Differences between Pb from the oil spill and other sources of Pb were determined through the investigation of Pb isotope composition. Important isotopic differences are observed between samples from the ships' hold and the natural background (Figure 4). Pb isotopic composition tends to be more radiogenic (high $^{206}Pb/^{207}Pb$ ratio) in BB sediments, which mimics the composition of oil collected inside the ship. It is important to note that BB is strongly impacted by the disposal of raw sewage, but the impact of this effluent is only apparent in the inner part of the bay (Barros et al., 2010). Samples in this investigation were collected towards the mouth and therefore should not be greatly influenced by raw sewage. The system indicates a simple binary mixing of geogenic with anthropogenic Pb.

Table 3. Lead concentration (mg/kg) in BB sediments

Site	Pb (mg/kg)
1	1.28
2	2.32
3	1.9
4	1.2
5	1.98
6	1.94
7	0.99
8	1.35
9	1.89

Lead released from the structures of silicate minerals is isotopically distinct from lead released from other sources, demonstrating distinct lead reservoirs. A different behavior can be observed among leachate and the residual phase of samples. Leachate presents a Pb isotopic composition, which is, in general, more radiogenic (marked by a $^{206}Pb/^{207}Pb$ ratio between 1.1465 and 1.1824) than the residual phase (marked by a $^{206}Pb/^{207}Pb$ ratio between 1.1052 and 1.1546). The sediment at the sampling locations had a $^{206}Pb/^{207}Pb$ ratio of 1.1766, providing clear evidence that they have an anthropogenic source. Meanwhile, lower ratios within the leachate suggest that no important anthropogenic contribution, such as sewage, was observed. A large amount of the anthropogenic lead added to sediment is most likely leached from the residue by dilute acid (Chow et al., 1973, Shirahata et al., 1980, Kober et al., 1999). The influence of organic matter (OM) and the iron oxyhydroxide surface can originate from reasonable differences between residual and leachate (Gioia et al. 2006). BB mangrove sediments have a great amount of OM. Another area of important anthropogenic contribution is in the vicinity of the Capri Iate Club; here the $^{206}Pb/^{207}Pb$ value is as low as 1.1362 from the residue. In the vicinity of the Sao Francisco do Sul Harbor, it is expected that the Pb isotopic composition would be close to that of the oil.

7.2. Carbon and Nitrogen

Carbon and nitrogen were analyzed following the same method previously presented for Pb. Unfortunately, sand beaches do not have sufficient OM, and, therefore, it was not possible to detect the carbon isotopic composition due to its low concentration. However, mangrove sediments were analyzed and the results are presented in table 2. Elemental and stable isotope analyses were used in a large number of studies to determine the spatial and temporal distribution of different sources of organic matter (allochthonous detritus and local phytoplankton) in suspended matter and sediments of estuarine systems – including BB (e.g. Cifuentes et al., 1996, Barros et al., 2010). This approach utilizes the fact that different biochemical pathways, and the degree to which carbon and nitrogen are reworked, will lead to different isotopic ratios of $\delta^{13}C$ and $\delta^{15}N$. $\delta^{13}C$ values range from -22.58‰ to -24.94‰. Heavier $\delta^{13}C$ values are observed at seaward sampling sites, suggesting a contribution of marine organic matter, which is characterized by a typical mean value -21.0‰ (Fisher et al. 1998). $\delta^{15}N$ values of BB sediments range from 3.56 to 6.43‰, with enriched values observed towards the ocean. Oil isotopic composition ranges from 8.24 to 8.64‰ for $\delta^{15}N$ and from -26.31 to -24.15‰ for $\delta^{13}C$. Sediment samples collected after the accident were marked by an isotopic composition varying from -24.69 to -2.07‰ for $\delta^{13}C$, and from 4.22 to 10.57‰ for $\delta^{15}N$. Just one organic matter sample matches with the oil sample. This discrepancy could be due to the time elapsed (nine months) between the accident and the fieldwork campaign. In addition several factors including tidal action alternatively emerging and submerging sample sites, two seasons (summer and autumn), and biogeochemical processes contributed to the isotopic fractionation.

CONCLUSION

Globally, estuaries are experiencing a strong loss of natural features due to recurrent acute impacts from human activities, and BB is certainly no exception. Herein, a forensic study was conducted to determine a unique source of pollution in an estuary that receives constant impacts from multiple sources. We utilized a multiple isotope analysis to help pinpoint a unique source of oil pollution in Babitonga Bay Carbon, nitrogen and lead isotopic ratios were used in conjunction with one another to better understand processes in BB. Carbon isotopes alone, would not have been able to tell the whole story. Instead, the use of isotopes, like Pb isotopes, which undergo little fractionation when exposed to weathering and biogeochemical processes is important and can be an interesting alternative, or complement, to carbon isotopes. Despite a relatively complex process, the use of combined isotopes tracers allowed for the possibility to spot areas where the oil spilled could be identified.

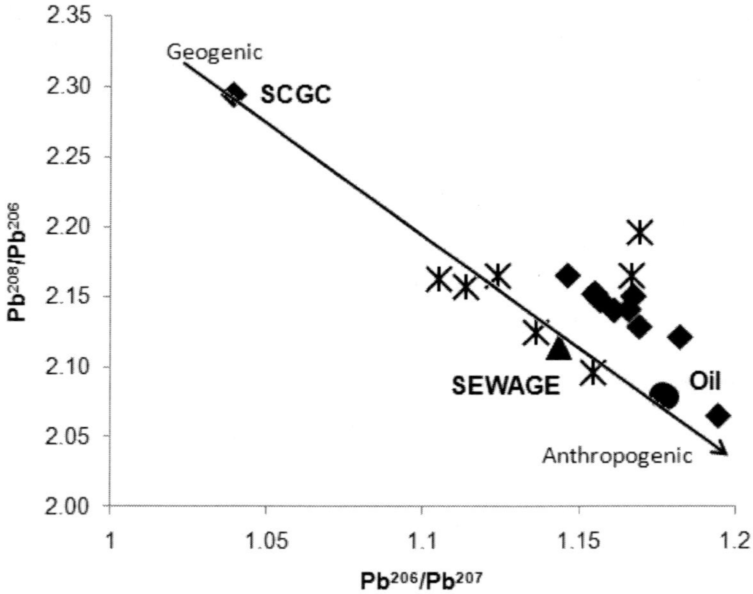

Figure 4. Isotopic differences between samples from the ships' hold and the natural background; a clear mixing line exists between the two.

Many of the world's most harmful pollutants contain lead. Pb isotopic compositions in Pb-containing products (e.g., gasoline, leaded pesticides, and leaded paints) depend only on the source ores; ores from different parts of the world often have their own Pb isotope signatures which are preserved during transport over vast distances and transfer processes. Therefore, from the ratios of the isotope concentrations of ^{208}Pb, ^{207}Pb, and ^{206}Pb detected in a sample, it is possible to pinpoint the source of Pb by matching the Pb isotopic composition found at the contamination site with those of the potential sources. It is also possible to apportion the contributions of the major sources based on their Pb isotopic compositions in cases where lead pollution resulted from multiple sources (Cheng and Hu, 2010). This study demonstrates the possibility and potential of the use of Pb isotopes to trace those responsible for environmental damage.

REFERENCES

Barros, G. V., Martinelli, L. A., Oliveira, N. T. M., Ometto, J. P. & Zuppi, G. M. (2010). Stable isotopes of bulk organic matter to trace carbon and nitrogen dynamics in an estuarine ecosystem in Babitonga Bay (Santa Catarina, Brazil) *Science of The Total Environm*, *408*, 2226-2232.

Barros, V. G. (2005). Study and monitoring of biogeochemical process of Babitonga Bay catchment—Brazil, using chemical and isotopes approaches. PhD Thesis, University of Ca' Foscari di Venezia, Venice, Italy.

Barros, V. G., Mas-Pla, J., Oliveira, T. M. N., Sacchi, E. & Zuppi, G. M. (2008). Hydrological mixing and geochemical processes characterization in na estuarine/mangrove system using environmental tracers in Babitonga Bay (Santa Catarina, Brazil). *Cont Shelf Research*, *28*, 682-695.

Carreira, R. S., Wagener, A. R., Fileman, T., Readman, J., Macko, S. A. & Veiga, A. (2002). Changes in sedimentary organic carbon pool of a fertilized tropical estuary, Guanabara Bay, Brazil: an elemental, isotopic and molecular marker approach. *Mar Chem*,*79*, 207–227.

Chow, T. J., Bruland, K. W., Bertine, K., Soutar, A., Koide, M. & Goldberg, E. D. (1973). Records in Southern California coastal sediments. *Science*, *81*, 551–552.

Cifuentes, L. A., Coffin, R. B., Solorzano, L., Cardenas, W., Espinoza, J. & Tilley, R. R. (1996). Isotopic and elemental variations of carbon and nitrogen in a mangrove estuary. *Estuar Coast-Shelf Sci.*, *43*, 781–800.

Clark, I. D. & Fritz, P. (1997). *Environmental Isotopes in Hydrogeology*, CRC Press, 328 p.

Coplen, T. B. (1994). Reporting of stable hydrogen, carbon, and oxygen isotopic abundances. *Pure Appl. Chem.*, *66*, 273 - 276.

Coplen, T. B. (1995). Discontinuance of SMOW and PDB. *Nature*, *375*(6529), 285.

Coplen, T. B. (1996). New guidelines for reporting stable hydrogen, carbon and oxygen isotope-ratio data. Geochim. et Cosmochim. *Acta*, *60*, 3359 - 3360.

Cremer, M. (Org). (2006). Diagnóstico ambiental da Baía da Babitonga. Joinville, SC, UNIVILLE

Dehairs, F., Rao, R. G., Chandra Mohan, P., Raman, A. V., Marguillier, S. & Hellings, L. (2000). Tracing mangrove carbon in suspended matter and aquatic fauna of the Gautami-Godavari Delta, Bay of Bengal (India). *Hydrobiologia*, *431*, 225–241.

Dittmar, T., Lara, R. J. & Kattner, G. (2001). River or mangrove?. Tracing major organic matter sources in tropical Brazilian coastal waters. *Mar Chem*, *73*, 253-271.

Fisher, T. R., Hagy, J. D. & Rochelle-Newall, E. (1998). Dissolved and particulate organic carbon in Chesapeake Bay. *Estuaries*, *21*, 215-229.

Gioia, S. M. C. L., Pimentel, M. M., Tessler, M., Dantas, E. L., Campos, J. E., Guimarães, E. M., Maruoka, M. T. & Nascimento, E. L. (2006). Sources of anthropogenic lead in sediments from an artificial lake in Brasília-central Brazil Science of The Total Environ, *356*, 125-142.

Gonfiantini, R. (1978). Standard for stable isotope measurement in natural compounds. *Nature*, *217*, 534-36.

Gonfiantini, R., Stichler, W. & Rozanski, K. (1995). Standards and intercomparison materials distributed by the International Atomic Energy Agency for stable isotope measurements. Reference and Intercomparison Materials for Stable Isotopes of Light Elements, IAEA-TECDOC-825. Vienna. IAEA.

Gonfiantini, R. (1978). Standard for stable isotope measurement in natural compounds. *Nature*, *217*, 534-536.

Goñi, M. A., Teixeira, M. J. & Perkeya, D. W. (2003). Sources and distribution of organic matter in a river-dominated estuary (Winyah Bay, SC, USA). *Estuar Coast Shelf Sci.*, *57*, 1023–1048.

Gonneea, M. E., Paytan, A. & Herrera-Silveira, J. A. (2004). Tracing organic matter sources and carbon burial in mangrove sediments over the past 160 years. *Estuar Coast Shelf Sci.*, *61*, 211-227.

Gordon, E. S. & Goñi, M. A. (2004). Controls on the distribution and accumulation of terrigenous organic matter in sediments from the Mississippi and Atchafalaya river margin. *Mar Chem*, *92*, 331-352.

Graham, M. C., Eaves, M. A., Farmer, J. G., Dobson, J. & Fallick, A. E. (2001). A study of carbon and nitrogen stable isotope and elemental ratios as potential indicators of source and fate of organic matter in sediments of the Forth Estuary, Scotland. *Estuar Coast Shelf Sci.*, *51*, 375–380.

Guadagnin, F., Chemale, Jr. F., Dussina, I. A., Jelineka, A. R., Santos, M. N., Borba, M. L., Justino, D., Bertotti, A. L. & Alessandretti, L. (2010). Depositional age and provenance of the Itajaí Basin, Santa Catarina State, Brazil: Implications for SW Gondwana correlation. *Precambrian Res.*, *180*, 156-182.

Guo, L., Tanaka, T., Wang, D., Tanaka, N. & Murata, A. (2004). Distributions, speciation and stable isotope composition of organic matter in the southeastern Bering Sea. *Mar Chem*, *91*, 211–226.

Hansson, S., Hobbie, J. E., Elmgren, R., Larsson, U., Fry, B. & Johansson, S. (1997). The stable nitrogen isotope ratio as a marker of food-web interactions and fish migration. *Ecology*, *78*, 2249-2257.

Hartmann, L. A., Santos, J. O. S., Mcnaughton, N. J., Vasconcellos, M. A. Z. & Silva, L. C. (2000). Ion Microprobe (SHRIMP) dates complex granulite from Santa Catarina, southern Brazil. *Anais da Acad Bras de Ciênc*, *72*, 559–572.

Hartmann, L. A., Silva, L. C. & Orlandi Filho, V. (1979). O Complexo Granulitico de Santa Catarina. Descrição e implicações genéticas. *Acta Geol. Leopoldensia*, *3*, 93-112.

Hefa, C. & Yuanan, H. (2010). Lead (Pb) isotopic fingerprinting and its applications in lead pollution studies in China: A review. *Environ Poll*, *158*, 1134–1146.

Hoda, A. & Häder, D. (2010). Rapid ecotoxicological bioassay of nickel and cadmium using motility and photosynthetic parameters of Euglena gracilis. *Environ and Exper Botany*, *69*, 68-75.

Hu, J., Peng, P., Jia, G., Mai, B. & Zhang, G. (2006). Distribution and sources of organic carbon, nitrogen and their isotopes in sediments of the subtropical Pearl River estuary and adjacent shelf, Southern China. *Mar Chem*, *98*, 274–285.

IBAMA (Instituto Brasileiro do Meio Ambiente). (1998). Proteção e controle de ecossistemas costeiros: manguezal da Baia de Babitonga. Instituto Brasileiro do Meio Ambiente e dos Recursos Naturais Renováveis. Coleção meio ambiente. Série estudos – pesca. Brazil.

Kendall, C. & McDonnell, J. J. (1998). Isotope Tracers in Catchment Hydrology. Amsterdam, Elsevier Science.

Kober, B., Wessels, M., Bollho, A. & Mangini, A. (1999). Pb isotopes in sediments of Lake Constance, central Europe constrain the heavy metal pathways and the pollution history of the catchment, the lake and the regional atmosphere. *Geochim Cosmochim Acta*, *63*(9), 1293–1303.

Komárek, M., Ettler, V., Chrastný, V. & Mihaljevič, M. (2008). Lead isotopes in environmental sciences: A review. *Environ Intern.*, *34*, 562–577.

Maksymowska, D., Richard, P., Piekarek-Jankowska, H. & Riera, P. (2000). Chemical and Isotopic Composition of the Organic Matter Sources in the Gulf of Gdansk (Southern Baltic Sea). *Estuar Coast Shelf Sci*, *51*, 585-598.

Mook, W. G. (Ed). (2001). Series on Environmental Isotopes in the Hydrological Cycle, Principles and Applications. UNESCO/IAEA, six volumes.

Oliveira, C. R. (2010). Speciation of butyltin derivatives in surface sediments of three southern Brazilian harbours. *Journal of Hazardous Materials*, *81*, 851-856.

Rutgers van der Loeff, M. M. (2001). Uranium-Thorium decay series in the oceans overview. In: Steel, J.H., Turekian, K.K., and Thorpe, S.A. (Editors), *Encyclopedia of Ocean Sciences*, 3135-3145.

Schettini, C. A. F., Carvalho, J. L. B. & Jabor, P. (1996). Comparative hydrology and suspended matter distribution of four estuaries in Santa Catarina State – Southern Brazil. *Proceedings of Workshop on Comparative Studies of Temperate Coast Estuaries*, 29-32.

Shirahata, H., Elias, R. W., Patterson, C. C. & Koide, M. (1980). Chronological variations in concentrations and isotope compositions of anthropogenic atmospheric lead in sediments of a remote subalpine pond. *Geochim Cosmochim Acta*, *44*, 149–162.

Thornton, S. F. & McManus, J. (1994). Applications of organic carbon and nitrogen stable isotope and C/N ratios as source indicators of organic matter provenance in estuarine systems: evidence from the Tay Estuary, Scotland. *Estuar Coast Shelf Sci*, *38*, 219–233.

USGS. (2004). Available at: http://wwwrcamnl.wr.usgs.gov/isoig/period/ pb_iig.html. (Acessed: 10th April 2014).

Vaz, C., Oliveira, T. M. N., Böhm, R. F. S., Spitzner, E. C., Simm, M. & Barros, V. G. (2010). Use of Artemia salina to identify sites with risk of contamination in the waters of Babitonga Bay. *Toxicology Letters*, *196*, S120.

Voss, M., Liskowa, I., Pastuszak, M., Rüβ, D., Schultec, U. & Dippnera, J. W. (2005). Riverine discharge into a coastal bay: A stable isotope study in the Gulf of Gdan´sk, *Baltic Sea. J Mar Syst.*, *57*, 127–145.

In: Oil Spills
Editor: Adam Clifton

ISBN: 978-1-63321-515-3

Chapter 7

USING WILD RODENTS AS A MODEL FOR PREDICTING IMPACTS OF ONSHORE OIL SPILLS

Flavio Manoel Rodrigues da Silva Júnior, Samantha Eslava Gonçalves Martins, Krissia Aparecida de Almeida, Fernanda Souza Mendonça, Paulo Roberto Martins Baisch and Ana Luíza Muccillo-Baisch

Laboratório de Ensaios Farmacológicos e Toxicológicos,
Instituto de Ciências Biológicas. Universidade Federal do Rio Grande - FURG,
Rio Grande, RS, Brasil

ABSTRACT

Crude oil is a mixture of many toxic compounds, including some known mutagens and metals such as polycyclic aromatic hydrocarbons and lead, respectively. The steps of exploration, onshore drilling, transportation and refining involve risk of accidental spills and may compromise environmental health and wildlife inhabiting the area. Among the animal species, free-living rodents play an important role either as seed dispersers or prey to many predators, since they represent intermediate links in the food web. These animals have been commonly used as indicators of disorders caused by environmental contaminants, especially metallic elements, and more recently organic contaminants, such as crude oil. We discuss the use of biomarkers in two species of free-living rodents exposed to simulated onshore oil spills, the Algerian mouse *Mus spretus* and the small Vesper mouse *Calomys laucha*. The studies indicate that genetic, hematological, physiological and behavioral biomarkers may be useful tools to predict stress situations facing a condition of an oil spill and point out those wild rodents could be used as risk indicators in natural populations exposed to this environmental contaminant.

Keywords: *Mus spretus*, *Calomys laucha*, petroleum, wild rodent

INTRODUCTION

The beginning of toxicological studies involving rodents is closely associated with the use of pharmacological assays in the preclinical screening of compounds with pharmacological potential. Important considerations such as route of administration, frequency of dose, vehicle, and administered volume are all concepts used by both Pharmacology and Toxicology. Moreover, Toxicology appropriated some concepts already used previously in Pharmacology, for example, drug metabolism, pharmacokinetics and pharmacodynamics. Within this context anchored by pharmacology, early studies with toxicological approach using laboratory rodents have emerged. Up to date, such studies are still useful for the understanding of the routes and mechanisms of toxic compounds, with the aim of using it in socio-environmental risk assessment [1]. Nevertheless, the studies performed with lab rodents have focused mostly on the effects of chemical substances alone or the interaction between a few compounds. Studies with complex environmental matrices, such as petroleum and its derivatives, are poorly addressed by the scientific community. Efforts should be made to intensify this field of research, since environmental exposures to these compounds involve the effects of the total mixture of the contaminants.

From the decade of 1960 on, the approach of using wild animals in biomedical research begun. The main reason for the use of these animals is because they share with human beings the vulnerability to several syndromes and diseases. In this sense, some species of wild rodents have contributed to clarify the mechanisms of some of these injuries [2].

Small wild mammals have been used to predict environmental impacts of metals since late 1970s [3,4]. In these early studies, trace metals accumulation was analyzed in different tissues in order to determine how far these contaminants could reach the interior of the animals, after environmental exposure to those compounds. Some rodent's features contribute to their suitability as a model for environmental contamination analysis, such as distribution, abundance, site fidelity, longevity and ecological representativeness [5].

More recently, studies have focused on toxicological aspects other than bioaccumulation, such as the toxic effects caused by the exposure to environmental pollutants, through the use of effect biomarkers. In this respect, studies conducted in order to elucidate genotoxic [6] and histopathological effects [7], antioxidant and metabolization enzymes [8], and damage to biomolecules [9] and immunological system [10] have been performed to investigate the toxicity in small wild mammals.

These wild organisms have also been used as models to examine toxic effects of trace metals under laboratory conditions [11-13]. Among the biomarkers applied for such analysis, there are Glutathione S-transferase (GST) activity (biotransformation enzyme), micronucleus induction, sister chromatid exchange, sperm abnormalities and plasma enzymes.

Currently, many species of small wild mammals have been used in environmental quality monitoring programs worldwide, with research conducted in Africa [14], North America [15], South America [6] and Europe [16]. Despite the use of wild rodents in many environmental studies, their utilization as sentinels in monitoring programs aiming to investigate impacts from oil spills and accidents in the terrestrial environment is still an incipient strategy.

Shubik et al. [17] conducted one of the first studies reporting toxic effects of grease oil. In this work, authors investigated three exposure routes: dermal, subcutaneous and dietary. Except for the dermal uptake of grease oil, which induced peeling, loss of hair and dermal hyperplasia the exposure to this contaminant did not change biological parameters significantly compared to the control group, in such a way that either exposed or non-exposed animals developed different kind of tumors.

Other studies regarding toxic effects of petrol and its derivatives were performed in laboratory strains of rodents, from late 1970s through the decade of 1980 [18-23]. Among the deleterious effects reported by these authors are clinical symptoms of acute toxicity such as tremors, paralysis and death, histopathological changes in the liver and kidneys, biochemical and hematological changes, and the appearance of tumors in chronically exposed animals. On the other hand, the oil fractions tested by Lockard and colleagues [21] showed weak mutagenicity and genotoxicity to somatic and germ cells.

In the following years, research has continued demonstrating the acute and chronic effects from the exposure to oil and oil products in lab rodents, through different routes of exposure. For example, Broddle and colleagues [24] studied the effects of dermal administration of different oil fractions for a long time (at least 24 months) and reported tumors development in the exposed animals. Adesanya et al. [25] examined the sperm toxic effects in Swiss mice exposed orally to Bonny light crude oil, showing reduced sperm density and motility in animals exposed to oil. Patrick-Iwuanyanwu et al. [26] verified hepatotoxic and nephrotoxic effects of food resources contaminated with oil and diesel oil in Wistar rats, and reported alterations in hepatic and renal biochemical markers, as well as degenerative changes in cells from these organs. Still in regard with dietary exposure to oil, Lemiere et al. [27] fed Wistar rats with contaminated shellfish collected from an oil spill area, resulting in an increase in genotoxic damage arising from the contaminated food consumption.

Studies involving wildlife are intended to investigate not only the modes of action related to toxic effects, but also to mark variables that respond early to an environmental disturbance. In this sense, research with contaminant bioaccumulation and effect biomarkers may provide with useful information in regard to contamination levels to which animals are exposed.

Schroeder and colleagues [28] investigated the pattern of inorganic compounds accumulation in bones and incisor teeth of the Cotton rat (*Sigmodon hispidus*) collected along three years in sites contaminated with substances related to petrochemical activities. Among the quantified elements, lead and fluoride accumulated more often and were related to the degree of soil contamination, whereas the seasonal variation in accumulation was associated to the ecological model of distribution and the life cycle (adult and young rats in winter and summer, respectively).

Genotoxicity and immunotoxicity markers have been applied to assess environmental risks associated to oil derivatives contaminants arising from anthropogenic activities. In Brazil, Heuser et al. [29] used two genotoxicity assays (comet assay and micronucleus test) in the native rodent *Ctenomys minutus* to monitor the influence of contaminants released by vehicular emissions on high traffic highway.

Authors concluded that the comet assay performed in peripheral blood was more efficient than the micronucleus test, and adult females demonstrated to be more affected by vehicular pollution. In Oklahoma, USA, Rafferty at al. [30] studied immunotoxic markers in the hispid Cotton rat (*S. hispidus*) in five sites contaminated with petrochemical waste. The most pronounced alterations in the animals obtained from contaminated sites were an increase in platelet count and peritoneal macrophages. Although other responses in the immune system were also affected in the animals coming from the contaminated areas, they were inconsistent when the five sites were compared.

Wilson and colleagues [31] adopted an interesting approach. They state that most of the studies on contamination coming from anthropogenic sources on wild animals, highlighting biomarkers research, focus on individual level, whereas little attention has been devoted to evaluate pollutant effects in ecological levels, such as population and communities. They investigated the differences in assemblies of wild rodents collected from abandoned petrochemical sites and non-contaminated sites. Results showed less diversity and species richness in contaminated areas, as well an increase in *Mus musculus* dominance and *Microtus* spp. absence.

In respect of the evaluation of adverse effects of onshore oil spills, the implementation of wild rodents as biological models is a strategy in early stages. However, some studies have already been performed using the Argelian mouse *Mus spretus* (Figure 1) and the small Vesper mouse *Calomys laucha* (Figure 2).

M. spretus inhabits the Iberian Peninsula, southern France and northern Africa. It is an omnivorous species that feeds both seeds and fruits as insect larvae and has interesting social structure, with males and females living in the same territory and high site fidelity [32].

The study conducted by Ieradi et al. [33] was one of the pioneers in using *M. spretus* as a key species in monitoring programs on environmental pollution and conservation. The authors analyzed chromosomal damage and micronucleus as biomarkers to compare the environmental quality status in an industrialized area and two non-contaminated area, one of them the Doñana National Park, in Spain. In the same year, this National Park was accosted by a disaster when a decanting dam in the Azalcollar pyrite mine broke, releasing tonnes of acid waters and mud with high trace metal concentrations. Following the accident, an emergential plan was carried for mud removal and water treatment, between May and October 1998. Based on previous data regarding the sensitivity of *M. spretus* to metals, several studies were performed to investigate the impacts of the disaster on Doñana National Park and to monitor this area, applying biochemical, genotoxic and molecular biomarkers [34-39].

At the same time, *M. spretus* was used as biomonitor species in active or inactive mining fields in Portugal [40-44]. Among the biomarkers used, there were dental characters, metallothionein levels, histopathological alterations, antioxidant enzymes activities, and morphological and hematological parameters.

Figure 1. *Mus spretus*.

Figure 2. *Calomys laucha*.

Studies using the small Vesper mouse *C. laucha* have also been conducted. *C. laucha* inhabits southern Bolivia, southern Brazil, central Argentina and Uruguay [45]. It is primarily granivorous and plays an important role in seed dispersion. This species has called the attention of the local scientific community and has been used as a biological model in research about trace metal effects. Worst [46] studied copper effects on reproduction and postnatal development, and Moraes [47] determined hematological parameters in animals exposed to copper contaminated drinking water.

These studies have highlighted the suitability of using *M. spretus* or *C. laucha* as sentinel organisms to predict the impacts arising from environmental pollution, including the consequences of environmental disasters. Da Silva Júnior et al. [48,49] evaluated the effects of simulated onshore oil spills in the Argelian mouse. In lab conditions, animals were kept on a substrate contaminated with Angolan crude oil (8% v/v), composed by fine sand (70%), mud (20%) and vegetal organic matter (10%). Mice were maintained in this artificial soil for 14 days, without substrate renewal. The exposure concentration is environmentally significant and followed that proposed by Gogoi et al. [50].

Drawing on a similar strategy, De Almeida [51], Da Silva [52] and Da Silva Júnior et al. [53] verified the effects of a simulated oil spill in health of the rodent *C. laucha*. In this study, Hydra light oil was used instead of Angolan crude oil, but the substrate and exposure conditions were the same described in the previous paragraph. Moreover, De Almeida [51] and Da Silva [52] investigated the toxic effects from a dose response curve (oil in the substrate: 1, 2, 4 and 8% v/v).

Many parameters were assessed by Da Silva Júnior et al. [48] in their study with *M. spretus*, including: clinical signs of acute toxicity, such as death and organ weight (liver, lung, kidney, testis, heart, and spleen); hematology; fecal corticosterone; oxygen consumption; locomotor activity and exploratory activity. On the other hand, the study conducted by Da Silva Júnior and colleagues [49] focused on genotoxicity and citotoxicity parameters, evaluated through the comet assay in peripheral blood cells, micronucleus test in bone marrow erythrocytes, and sperm abnormalities.

About studies conducted with *C. laucha* to predict oil spills damage, the parameters analyzed were: hematological alterations [53]; clinical signs of acute toxicity, such as food intake, paw volume, and organ weight (liver, lung, kidney, testis, heart, and spleen);

mutagenicity and citotoxicity in bone marrow erythrocytes; sperm viability and abnormalities; locomotor and exploratory activities [51]; and lipid damage, protein damage and alteration in antioxidant capacity against peroxyl radicals [52].

The rodents from both species showed similar acute intoxication signs, such as bristling hair, hair loss in the ventral portion, lethargy and limited mobility, beyond death of some organisms. However, significant weight loss was not seen in any of rodents (Table 1).

Table 1. Presence of external signs of acute toxicity in rodents exposed to a simulated oil spill

Parameter	*M. spretus*[1]	*C. laucha*[2]*
Bristling hair	+	+
Alopecia	+	+
Lethargy	+	+
Limited mobility	+	+
Weight loss	NO	NO
Death	+	+

*taking into account only the highest concentration analyzed - 8%(v/v), after 14-days exposure
[1] Extracted from Da Silva Júnior et al. [48]
[2] Extracted from De Almeida [51]
+ observed effect; NO non observed effect.

Table 2. Comparison of changes in the weight of the principal organs of rodents exposed to a simulated oil spill

Organ	*M. spretus*[1]	*C. laucha*[2]*
Liver	↑	↑
Spleen	↑[3]	↑
Lung	↓	=
Heart	=	=
Right kidney	=	↑
Left kidney	=	↑
Right testicle	=	↓
Left testicle	=	=

*taking into account only the highest concentration analyzed – 8% (v/v).
[1] Extracted from Da Silva Júnior et al. [48].
[2] Extracted from De Almeida [51].
[3] significant increase of the spleen width.

Although body weight did not seem to be affected after oil exposure in a simulated spill (Table 1), measures of internal organs were effective parameters to indicate damage from exposure to oil (Table 2). Despite this information, only liver, spleen, heart and left testicle showed similar responses between the two species. Liver and spleen were increased while heart and left testicle were not altered by oil exposure in both species. In the other organs analyzed (lung, kidney and right testicle), the responses to oil exposure were different in each species.

Conventional hematological parameters were not affected by oil exposure in *M. spretus*, but *C. laucha* demonstrated high sensitivity to detect changes in some of these parameters (Table 3), such as an increase in white blood cells related to the establishment of an acute inflammatory process; and increase in platelet count related to the stimulation of megacariocytosis in bone marrow.

Table 3. Comparison of changes in hematological profile of rodents exposed to a simulated oil spill

Parameter	*M. spretus*[1]	*C. laucha*[2]*
Erythrocytes	=	=
Hemoglobin	=	=
Hematocrit	=	=
MVC	=	=
MCH	=	=
MCHC	=	=
White blood cells	=	↑
Platelets	=	↑

*taking into account only the highest concentration analyzed – 8% (v/v).
[1] Extracted from Da Silva Júnior et al. [53].
[2] Extracted from De Almeida [51].

Studies with these two species were the first to evaluate alterations in wild rodents through behavior tests, regarding neurotoxic effects of environmental contaminants. The apparatus used (open field test) is capable of measuring changes in locomotor and exploratory activity, and further suggests behavior related to fear and anxiety. Table 4 shows the results comparing the two species. In the Algerian mouse, oil exposure diminished locomotor activity (peripheral crossing) but stimulated fear and anxiety (grooming and freezing). In the small Vesper mouse, oil exposure was primarily related to the reduction of the exploratory activity.

Despite the differences between the responses in different species, this test proves to be a promising tool for the evaluation of neuro-behavioral changes associated to the exposure of small wild mammals to environmental contaminants.

Table 4. Comparison of changes in behavioral parameters measured by the open field test in rodents exposed to a simulated oil spill

Parameter	*M. spretus*[1]	*C. laucha*[2]*
Peripheral Crossing	↓	=
Central Crossing	=	↓
Rearing	=	↓
Grooming	↓	=
Freezing	↑	=

*taking into account only the highest concentration analyzed - 8%(v/v), after 14-days exposure.
[1] Extracted from Da Silva Júnior et al. [48].
[2] Extracted from De Almeida [51].

In regard of physiological parameters, exposure to simulated oil spill affected both species, *M. spretus* and *C. laucha*. Although the analyzed parameters were different between the two species, oil seems to change the general metabolism of rodents (Table 5). These measurements are minimally invasive and non-destructive parameters, eliminating the euthanasia of the animal. In this sense, their use is advantageous, and they can be applied as an alternative tool in monitoring programs concerning environmental disasters, along with other biological measurements.

Table 5. Comparison of physiological parameters in rodents exposed to a simulated oil spill

Parameter	*M. spretus*[1]	*C. laucha*[2*]
Fecal corticosterone	↓	NA
Oxygen consumption	↑	NA
Food intake	NA	↓
Muscle weight	NA	↓

*taking into account only the highest concentration analyzed - 8%(v/v), after 14-days exposure.
[1] Extracted from Da Silva Júnior et al. [48].
[2] Extracted from De Almeida [51].
NA non analyzed parameter.

Mutagenicity, genotoxicity and citotoxicity biomarkers used to assess negative effects of oil exposure in wild rodents showed to be extremely sensitives (Table 6). Bioassays were conducted using cells from different lines (leucocytes, erythrocytes and sperm cells) and every cell type was damaged by substances present in the oil. In this sense, the high sensitivity to geno and citotoxic compounds makes the Algerian mouse and the small Vesper mouse suitable as biomonitor species of genetic risks of the wild population to oil in environmental disasters situations.

Table 6. Comparison of genotoxicity, mutagenicity and citotoxicity parameters in rodents exposed to a simulated oil spill

Parameter	*M. spretus*[1]	*C. laucha*[2*]
DNA damage (comet assay)	+	NA
Bone marrow (mutagenicity)	+	+
Bone marrow (citotoxicity)	+	+
Sperm abnormalities	+	+
Sperm viability	NA	+

*taking into account only the highest concentration analyzed - 8%(v/v), after 14-days exposure.
[1] Extracted from Da Silva Júnior et al. [49].
[2] Extracted from De Almeida [51].
NA non analyzed parameter.

Finally, measurements related to the oxidative status revealed that when *C. laucha* was exposed to oil, this rodent increased the antioxidant capacity against peroxyl radical, however it was unable to avoid lipid and protein damage (Table 7). This pro-oxidative status was observed in the liver. Although it might be reasonable to believe that this profile also occurs in other organs, it is known that this kind of biochemical response is strictly tissue- or organ-dependent, so studies in other organs are necessary.

Table 7. Comparison of parameters related to oxidative stress in rodents exposed to a simulated oil spill

Parameter	*C. laucha*[1]*
Lipid damage	↑
Protein damage	↑
Antioxidant capacity	↑

*taking into account only the highest concentration analyzed - 8%(v/v), after 14-days exposure.
[1] Extracted from Da Silva [52].

FINAL REMARKS

Based on the recent studies with simulated oil spills using two rodent species (*M. spretus* and *C. laucha*), a variety of parameters or biomarkers that substantially meet the exposure to the mix of contaminants can be indicated. Signs of acute intoxication, and biochemical, genotoxic, hematological, physiological and behavioral markers were altered by exposure to oil and can be applied in monitoring studies of impacts from accidental oil spills in different species of wild rodents. In addition, other parameters already used for other species can be aggregated for studies of disaster monitoring.

Due to the easiness of sample collection, we suggest a sequence of parameters to be used in these studies (Figure 3). Initially, parameters that do not require the capture of live animals, even temporarily, should be prioritized. These include samples from corpses, feces, hair, teeth, and all associated biomarkers, as well as observations of significant changes in animal behavior or signs of acute toxicity still in the field.

Then, markers that require temporary capture of the animal but without a high degree of discomfort can be used. These markers include morphometry, body weight, photographic records and hair collection.

The third group includes the parameters that need a certain degree of invasiveness, but removal of the animal from the field to the laboratory is not required. Parameters of this group need administration of anesthetics, and comprise blood collection, nail samples and measures of muscle mass through the paw volume, among others.

Some measurements are non-destructive, but should be performed in the laboratory. Ideally set up a laboratory in the field itself, to accomplish the release of animals immediately after the measurements. Behavioral parameters, oxygen consumption, and food and drink uptake measurements are included in this group.

Finally, the fifth group comprises measures that require euthanasia of the animal and subsequent autopsy for removal of organs and tissues. Belong to this group, studies of the distribution of contaminants in different organs, and their response after oil exposure.

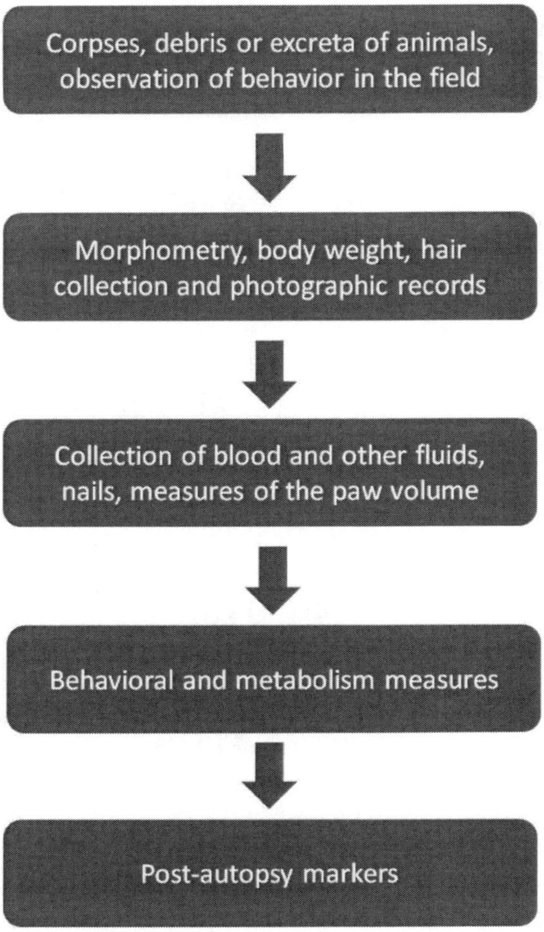

Figure 3. Sample and parameters chain investigated for predicting damage from oil spills in wild rodents.

CONCLUSION

This chapter has demonstrated the importance of the use of wild rodents as a model to predict the impacts of onshore oil spills. The studies already performed have demonstrated different ways in which rodents can be used to evaluate pollution effects. In all cases, results showed that these animals are sensitive to oil exposure.

When biological models are used, the modes of action of toxic substances can be understood. In this sense, the early response to environmental disturbances can be detected through the use of biomarkers.

Many biomarkers are useful for determining effects of oil and its derivatives, making the wild rodents suitable for use in monitoring programs and environmental risk assessment. Taking into account all methodologies and results described in this chapter, it has highlighted the suitability of using *M. spretus* or *C. laucha* as sentinel organisms to predict the impacts arising from environmental pollution, including the consequences of environmental disasters.

REFERENCES

[1] Hamm, TE; Vasbinder, MA; King-Herbert, A. Toxicology. In: Suckow, MA; Weisbroth, SH; Franklin, CL. The Laboratory rat. Burlington: Elsevier; 2006; p. 803-816.

[2] Hulin, M.; Quinn, R. Wild and Black rats. In: Suckow, MA; Weisbroth, SH; Franklin, CL The Laboratory rat. Burlington: Elsevier; 2006; p. 865-882.

[3] Beardsley, A; Vagg, MJ; Beckett, PHT; Sansom, BF. Use of the field vole (M. agrestis) for monitoring potentially harmful elements in the environment. *Environmental Pollution*, 1978, v. 16, p. 65-71.

[4] Roberts, RD; Johnson, MS. Lead contamination of small mammals from abandoned metalliferous mines. *Environmnetal Pollution*, 1978, v. 15, p. 61-69.

[5] Tataruch, F; Kierdorf, H. Mammals as biomonitors. In: Markert, BA; Breure, AM; Zechmeister, HG. Bioindicators and biomonitors. Burlington: Elsevier Science; 2003; p. 737-772.

[6] Da Silva, J; Freitas, TRO; Heuser, V; Marinho, JR; Bittencourt, F; Cerski, CTS; Kliemann, LM; Edrtmann, B. Effects of chronic exposure to coal in wild rodents (*Ctenomys torquatus*) evaluated by multiple methods and tissues. *Mutation Research*, 2000, v. 470, p. 39-51.

[7] Kim KB, Lee BM. Oxidative stress to DNA, protein, and antioxidant enzymes (superoxide dismutase and catalase) in rats treated with benzo(a)pyrene. *Cancer Letters*, 1997; v. 113; p. 205-212.

[8] Fouchécourt, M; Rivière, J. Activities of cytochrome P450-dependent monooxygenases and antioxidant enzymes in different organs of norway rats (*rattus norvegicus*) inhabiting reference and contaminated sites. *Chemosphere*, 1995, v. 31, p. 4375-4386.

[9] Baya-Rodriguez, MT; Ferreira, CJS; Moreira, JCF. Impacto automotivo em populações de Ctenomys minutus na planície costeria do RS; avaliação do teor de metais tóxicos e medida de lipoperoxidação. *Journal of the Brazilian Society of Ecotoxicology*, 2006, v. 1, p.177-183.

[10] Propst, TL; Lochmiller, RL; Qualls, CW; Mcbee, K. In situ (mesocosm) assessment of immunotoxicity risks to small mammals inhabiting petrochemical waste sites. *Chemosphere,* 1999, v. 38, p. 1049-1067.

[11] Pinheiro, T; Rocha, P; Lopes, PA; Viegas-Crespo, AN, Santos, MC; Mathias, ML. Laboratory testing of biochemical effects of Mn, Fe and Se in the Algerian mouse (Mus spretus). *Ecotoxicology and Environmental Restoration*, 2001, v. 4, p. 60-63.

[12] Marques, CC; Nunes, AC; Pinheiro, T; Lopes, PA; Santos, MC; Viegas-Crespo, AM; Ramalhinho, MG; Mathias, ML. An assessment of time-dependent effects of lead exposure in algerian mice (*Mus spretus*) using different methodological approaches. *Biological Trace Element Research,* 2006, v. 109, p. 75-89.

[13] Tapisso, JT; Marques, CC; Mathias, ML; Ramalhinho, MG. Induction of micronuclei and sister chromatid exchange in bone-marrow cells and abnormalities in sperm of Algerian mice (*Mus spretus*) exposed to cadmium, lead and zinc. *Mutation Research*, 2009, v. 678, p. 59-64.

[14] Nakayama, SM; Ikenaka, Y; Hamada, K; Muzandu, K; Choongo, K; Yabe, J; Umemura, T; Ishizuka M. Accumulation and biological effects of metals in wild rats in

mining areas of Zambia. *Environmental Monitoring Assessment*, 2013, v. 185, p.4907-4918.

[15] Espinosa-Reyes, G; Torres-Dosal, A; Ilizaliturri, C; Gonzalez-Mille, D; Diaz-Barriga, F; Mejia-Saavedra, J. Wild rodents (*Dipodomys merriami*) used as biomonitors in contaminated mining sites. *Journal of Environmental Science and Health Part A*, 2010, v. 45, p. 82–89.

[16] Gonzalez, XI; Aboal, JR; Fernandez, JA; Carballeira, A. Evaluation of some sources of variability in using small mammals as pollution biomonitors. *Chemosphere*, 2008, v. 71, p. 2060–2067.

[17] Shubik, P; Saffiotti, U; Lijinsky, W; Fetra, G.;. RaPaport, H.; Toth, B; Raha, CR; Tomatis, L; Feldman, R; Raiuaiix, H. Studies on the Toxicity of Petroleum Waxes. *Toxicology and Applied Pharmacology*, 1962, v. 4, p. 1-62.

[18] Veldre, A; Janes, HJ. Toxicological Studies of Shale Oils, Some of Their Components, and Commercial Products. *Environmental Health Perspectives,* 1979, v. 30, p. 141-146.

[19] Parker, GA; Bogo, V; Young, RW. Acute Toxicity of Conventional versus Shale-Derived JP5 Jet Fuel: Light Microscopic, Hematologic, and Serum Chemistry Studies. *Toxicology and Applied Pharmacology*, 1981, v. 57, p. 302-317.

[20] Easley, JR; Holland, JM; Gipson, LC; Whitaker, MJ. Renal toxicity of middle distillates of shale oil and petroleum in mice. *Toxicology and Applied Pharmacology*, 1982, v. 65, p. 84-91.

[21] Lockard, JM; Prater, JW; Viau, CJ; Enoch, HG; Sabharwal, PS; Hansberger, NS; Gleason, JR; Kamber, SJ; Lambert, CE; Carter, JC. Comparative study of the genotoxic properties of Eastern and Western U.S shale oils, crude petroleum, and coal-derived oil. *Mutation Research*, 1982, v. 102, p. 221-235.

[22] Khan, AA; Coppock, RW; Schuler, MM; Geleta, L. Biochemical changes as early stage systemic biomarkers of petroleum hydrocarbon exposure in rats. *Toxicology Letters*, 2002, v. 134, p.195-202.

[23] Wilson, JA; Lochmiller, RL; Janz, DM. Dynamics of rodent assemblages inhabiting abandoned petroleum landfarms in Oklahoma. *Ecological Applications*, 2004, v. 14(4), p. 1016–1027.

[24] Broddle, WD; Dennis, MW; Kjtchen, DN; Vernot, EH. Chronic Dermal Studies of Petroleum Streams in Mice. *Fundamental and Applied Toxicology,* 1996, v. 30, p. 47-54.

[25] Adesanya, OA; Shittu, LA; Omonigbehin, EA; Tayo, AO. Spermatotoxic impact of bonny light crude oil (BLCO) ingestion on adult male Swiss albino mice. *International Journal of Physical Sciences*, 2009, v. 4 (5), p. 349-353.

[26] Patrick-Iwauanyanwu, KC; Onyemaenu, CC; Wegwu, MO; Ayalogu, EO. Hepatotoxic and nephrotoxic effects of kerosene and petrol-contaminated diets in wistar albino rats. *Research Journal of Environmental Toxicology,* 2011, v. 5, p. 49-57.

[27] Lemiere, S; Cossu-Leguille, C; Bispo, A; Jourdain, M; Lanhers, M; Burnel, D; Vasseur, P. DNA damage measured the single-cell gel electrophoresis (Comet) assay in mammals fed with mussles contaminated by the 'Erika' oil-spill. *Mutation Research*, 2005, v. 581, p. 11-21.

[28] Schroder, JL; Basta, NT; Lochmiller, RL; Rafferty, DP; Payton, M; Kim, S; Qualls Jr. CW. Soil contamination and bioaccumulation of inorganics on petrochemical sites. *Environmental Toxicology and Chemistry*, 2000, v. 19, p. 2066–2072.

[29] Heuser, VD; Silva, J; Moriske, H; Dias, JF; Yoneama, ML; Freitas, TRO. Genotoxicity Biomonitoring in regions exposed to vehicle emissions using the comet assay and the micronucleus test in native rodent *Ctenomys minutes Environmental and Molecular Mutagenesis*, 2002, v. 40, p. 227–235.

[30] Rafferty, DP; Lochmiller, RL; Qualls, CW. Immunotoxicity risks associated with land-treatment of petrochemical wastes revealed using an in situ rodent model. *Environmental Pollution*, 2001, v. 112, p. 73-87.

[31] Wilson, JS; Holland, LM. Periodic response difference in mouse edipermis chronically exposed to crude-oils or B(a)p: males vs females. *Toxicology,* 1988, v. 50, p. 83-94.

[32] Palomo, LJ; Justo, ER; Vargas, JM. *Mus spretus* (Rodentia: Muridae). *Mammalian species*, 2009, v. 840, p. 1-10.

[33] Ieradi, LA; Moreno, S; Bolívar, JP; Cappai, A; Di Benedetto, A; Cristaldi, M. Free-living rodents as bioindicators of genetic risk in natural protected areas. *Environmental Pollution*, 1998, v. 102, p. 265-268.

[34] Tanzarella, C; Degrassi, F; Cristaldi, M; Moreno, S; Lascialfari, A; Chiuchiarelli, G; Ieradi, LA. Genotoxic damage in free-living Algerian mouse (*Mus spretus*) after the Coto Doñana ecological disaster. *Environmental Pollution*, 2001, v. 115, p. 43-48.

[35] Festa, F; Cristaldi, M; Ieradi, LA; Moreno, S; Cozzi, R. The comet assay for the dection of DNA damage in *Mus spretus* from Doñana National Park. *Environmental Research*, 2003, v. 91, p. 54-61.

[36] Bonilla-Valverde, D; Ruiz-Laguna, J; Munõz, A; Ballesteros, J; Lorenzo, F; Gómez-Ariza, JL; López-Barea, J. Evolution of biological effects of Aznalcóllar mining spill in the Algerian mouse (*Mus spretus*) using biochemical biomarkers. *Toxicology,* 2004, v. 197, p. 123-138.

[37] Montes-Nieto, R; Fuentes-Almargo, CA; Bonilla-Valverde, D; Prieto-Álamo, MJ; Jurado, J; Carrascal, J; Gómez-Ariza, JL; López-Barea, J; Puyeo, C. Proteomics in free-living Mus spretus to monitor terrestrial ecosystems. *Proteomics,* 2007, v. 7, p. 4376-4387.

[38] Leoni, S; Torrice, A; Chiuchiarelli, G; Bruscalupi, G; Cristaldi, M. Cell-Cycle Protein Modifications in the Kidney of *Mus spretus* from Doñana National Park. *Environmental Toxicology*, 2008, v. 23, p. 44-51.

[39] Mateos, S; Daza, P; Domínguez, I; Cárdenas, JA; Cortés, F. Genotoxicity detected in wild mice living in a highly polluted wetland area in south western Spain. *Environmental Pollution*, 2008, v. 53, p. 590-593.

[40] Nunes, AC; Auffray, JC; Mathias, ML. Developmental instability in a riparian population of the Algerian Mouse (*Mus spretus*) Associated with a heavy metal-polluted area in central Portugal. *Archives of Environmental Contamination and Toxicology*, 2001, v. 41, p. 515-521.

[41] Nunes, AC; Mathias, ML; Crespo, AM. Morphological and haematological parameters in the Algerian mouse (*Mus spretus*) inhabiting an area contaminated with heavy metals. *Environmental Pollution,* 2001, v. 113, p. 87-93.

[42] Viegas-Crespo, AM; Lopes, PA; Pinheiro, MT; Santos, MC; Rodrigues, PD; Nunes, AC; Marques, C; Mathias, ML. Hepatic elemental contents and antioxidant enzymes activities in Algerian mouse (*Mus spretus*) inhabiting a mine area in central portugal. *Science of the Total Environment,* 2003, v. 311, p. 101-109.

[43] Pereira, R; Pereira, ML; Ribeiro, R; Gonçalves, F. Tissues and hair residues and histopathology in wild rats (*Rattus rattus* L) and Algerian mice (*Mus spretus* Lataste) from and abandoned mine area (Southeast Portugal). *Environmental Pollution*, 2006, v. 139, p. 561-575.

[44] Marques, CC; Gabriel, SI; Pinheiro, T; Viegas-Crespo, AM; Mathias, ML; Bebianno, MJ. Metallothionein levels in Algerian mice (*Mus spretus*) exposed to elemental pollution: An ecophysiological approach. *Chemosphere,* 2008, v. 71, p. 1340-1347.

[45] Nowack, RM; Paradiso, J. Walker`s Mammals of the World. 4th ed., Baltimore: John Hopkins University Press; 1983;1362 pp.

[46] Worst, MR. Efeitos da exposição ao cobre no desenvolvimento pós-Natal e na reprodução de fêmeas de *Calomys laucha* Olfers, 1818 (Rodentia: Muridae). Dissertação de Mestrado. Universidade Federal do Rio Grande. Programa de Pós-Graduação em Ciências Fisiológicas - Fisiologia Animal Comparada. Rio Grande. 2011.

[47] Moraes, LN. Efeitos da exposição ao cobre pela água de consumo nos parametros hematológicos e na acumulação do metal no sangue e fígado em *Calomys laucha* Olfers, 1818 (Mammalia: Rodentia). Dissertação de (Mestrado). Universidade Federal do Rio Grande. Programa de Pós-Graduação em Ciências Fisiológicas - Fisiologia Animal Comparada. Rio Grande. 2011.

[48] Da Silva-Júnior, FMR; Monarca, RI; Dias, D; Ramalhinho, MG; Mathias, MM; Muccillo-Baisch, AL. Physiological damage in Algerian mouse *Mus spretus* (Rodentia, Muridae) exposed to crude oil. *Journal of BioScience and Biotechnology*, 2012, v. 1, p. 125-133.

[49] Da Silva-Júnior, FMR; Monarca, RI; Dias, D; Ramalhinho, MG; Mathias, MM; Muccillo-Baisch, AL. Geno- and Cyto-toxicity in Free-Living Rodent *Mus spretus* Exposed to Simulated Onshore Oil Spill. *Bulletin of Environmental Contamination and Toxicology*, 2013, v. 91, p. 465-468.

[50] Gogoi BK, Ditto NN, Goswami P, Krishna Mohan TR. A case study of bioremediation of petroleum-hydrocarbon contaminated soil at a crude oil spill site. *Advances in Environmental Research*, 2003, v. 7, p. 767-782.

[51] De Almeida, KA. Avaliação da toxicidade de solos contaminados por petróleo através de biomarcadores fisiológicos em Calomys laucha Olfers 1818. Monografia de Conclusão de Curso – Tecnologia em Toxicologia Ambiental, FURG. Rio Grande. 2012.

[52] Da Silva, P. Investigação de danos oxidativos relacionados à exposição de roedores a solo contaminado por petróleo. Monografia de Conclusão de Curso – Tecnologia em Toxicologia Ambiental, FURG. Rio Grande. 2012.

[53] Da Silva-Júnior, FMR; De Almeida, KA; Silva, PF; Muccillo-Baisch, AL. Hematological profile as a crude oil exposure-related marker in wild rodent. *Journal of BioScience and Biotechnology*, 2013, v. 2, p. 89-94.

In: Oil Spills
Editor: Adam Clifton

ISBN: 978-1-63321-515-3

Chapter 8

BIOREMEDIATION OF CRUDE OIL SPILLS BY INDIGENOUS MICROORGANISMS

Xiaoke Hu[1,*], Meng Lin[1], Xin Liu[2] and Chuanyuan Wang[2]

[1]Key Laboratory of Coastal Biology and Resource Utilization, Yantai Institute of Coastal Zone Research, Chinese Academy of Sciences, Yantai, P.R. China

[2]Key Laboratory of Coastal Zone Environmental Processes, CAS; Shandong Provincial Key Laboratory of Coastal Zone Environmental Processes; Yantai Institute of Coastal Zone Research, Chinese Academy of Sciences, Yantai, P. R. China

ABSTRACT

Oil spills occur either in the ocean and coastal waters or on land due to the release of crude oil from tankers, offshore platforms, drilling rigs and wells. It is becoming a major widespread environmental problem, especially in coastal areas subjected to high anthropogenic inputs. The high crude oil concentrations accumulated in surface and subsurface of ocean and soil pose serious concerns for the environmental and human health. Bioremediation, mainly by indigenous bacteria, has been regarded as an effective way to clean up oil pollution after oil spills. In order to obtain a systematic understanding of crude oil bioremediation, two biodegradation approaches, bio-stimulation and bio-augmentation will be summarized. Moreover, the detailed studies of the Deepwater Horizon oil spill in the Gulf of Mexico and the Penglai 19-3 oil spill in the Bohai Bay, China will be compared. The immobilization techniques used in the bioremediation process will also be reviewed. The insights and theories of these studies will improve our understanding of the microbial degradation process and mechanisms of indigenous microorganisms and provide fundamental evidences and strategies for *in situ* bioremediation of crude oil spills in the marine ecosystem.

INTRODUCTION

Crude oil is toxic for most life beings. Oil spills refer to petroleum hydrocarbons naturally or accidentally released to the environment and can usually lead to long-term negative impacts. Episodic and chronic pollution of the environment by oil causes major ecological perturbations. Marine environments are especially vulnerable since oil spills of coastal regions and the open sea are poorly containable and mitigation is difficult. In addition to pollution through human activities, about 250 million liters of petroleum enter the marine environment each year from natural seepages (MacDonald et al., 2002). Since the oil spills affect marine life dramatically, the impact of oil spill on the ecosystem is severe and could not be over-emphasized. The ability of marine ecosystems to recuperate from oil contamination is largely dependent on the composition and activities of indigenous hydrocarbon degrading microorganisms, which varies in different ecosystem in question. A considerable fraction of crude oil entering marine systems is eliminated by the hydrocarbon-degrading activities of microbial communities, in particular by a remarkable recently discovered group of specialists, the so-called hydrocarbonoclastic bacteria (Yakimov et al., 2007). *Alcanivorax borkumensis* was the first hydrocarbonoclastic bacteria to have its genome sequenced (Martins dos Santos et al., 2008). In addition to hydrocarbons, crude oil often contains various heterocyclic compounds, such as pyridine, which appear to be degraded by separate mechanisms than hydrocarbons.

The process of bioremediation is for microorganisms to break up complex compounds into smaller compounds for their food, detoxify or remove pollutants through their diverse metabolic capabilities (Medina-Bellver et al., 2005). This process usually occurs naturally by indigenous microorganism communities. The clean-up of oil spills involves the artificial introduction of biological agents such as fertilizers and nutrients to indigenous microorganisms *in situ* to stimulate their proliferation (bio-stimulation) or the introduction of non-native microorganisms (bio-augmentation) to speed up the natural process of biodegradation (Al-Majed et al., 2012). Bioremediation technology is believed to be a non-invasive, relatively cost-effective and cheaper than other remediation technologies (April et al., 2000; Leahy and Colwell, 1990).

INDIGENOUS MICROORGANISMS PLAY IMPORTANT ROLES IN BIOREMEDIATION

As one of the largest marine oil pollutions, the Deepwater Horizon oil spill released approximately 4.9 million barrels of crude oil from April 20 to July 15, 2010. Around 2 million gallons of chemical dispersants Corexit were applied to the sea water to improve the bioavailability of crude oil and speed up the bioremediation process (Crone and Tolstoy,2010; Joye et al., 2011). The large volumes of oil released in the extreme depth between 1000 m and 1300 m of the blowout made the effectiveness of the primary initial mitigation strategy (e.g., injecting the oil dispersant Corexit 9500 directly at the wellhead in a water depth of 1544 m) difficult to apply. Therefore, using of the indigenous deep-sea microorganisms to eliminate the crude oil pollution becomes one optional strategy. Hazen et al. (2010) reported the first research paper in Science, demonstrated that the dispersed hydrocarbon plume stimulated deep-sea indigenous γ-Proteobacteria. The hydrocarbon degradation rate was faster than expected at a lower temperature (5 °C). Members of the *Oceanospirillales* and the genus *Halomonas* were reported to be dominant in the deep sea hydrocarbon plume (Hazen et al., 2010). The deep water plume microbial community responded to the Deepwater Horizon oil spill was found to be distinct from the observed in previous spills or mesocosm studies. Valentine's group reported in PNAS (2011) that the microbial communities present in the Gulf of Mexico were initially dominated by members of *Oceanospirillales*, *Colwellia*, and *Cycloclasticus*, which were not abundant in surface oil slick samples. The colder temperature at the plume depth favoured to develop these specific bacterial communities. The abundance of the natural gas in the Deepwater Horizon spill seemed to provide an advantage to *Colwellia* and *Oceanospirillales*, because both of them became more abundant with the addition of ethane or propane alone. Firmicutes and α-Proteobacteria were found in open ocean surface slicks. The analogy study demonstrated that large-scale release of methane from gas hydrate in the deep ocean is likely to be completely oxidized by those bacteria from a rapid methanotrophic response (Kessler et al., 2011). A significant amount of spilled oil from the Deepwater Horizon oil spill was transported to the shoreline. Kostka et al. (2011) identified the predominant oil-degrading taxa and characterized the *in situ* response of indigenous bacterial communities to oil pollution in beach ecosystems. Their study revealed that members of the γ-Proteobacteria (*Alcanivorax*, *Marinobacter*), and α-Proteobacteria (*Rhodobacteraceae*) were dominant in coastal beaches after the Deepwater Horizon oil spill.

Bohai Bay is the largest oil and gas production base offshore China. The Penglai 19-3 oil spill occurred in the Bohai Bay of China on June 4, 2011. It released large amounts of crude oil (equivalent of 1500 barrels of oil) into the Bohai Bay. Our group (Wang et al., 2013) investigated the phylogenetic diversity of bacterial communities associated with biodegradation of crude oil in microcosm tests. Our research suggested that crude oil, especially alkanes, was efficiently biodegraded. Efficient bioremediation was related to the dramatic changes of bacterial communities before and after oil stress, particularly the variability of so-called 'professional hydrocarbonoclastic bacteria' - *Alcanivorax* emerging and becoming more abundant in samples with degraded crude oil. Bacteria affiliated with δ-Proteobacteria, especially in the genus *Achromobacter* were also detected in these two samples. Moreover, there were no dramatic differences in either the oil degrading rate or bacterial communities with the addition of the surfactant Tween-20. We believed that the hydrocarbon-degrading bacteria have the capability of producing biosurfactant themselves to disperse the crude oil.

OIL WEATHERING PROCESS

Once crude oil has entered into the environment, it is subjected to a variety of weathering processes, such as evaporation, dissolution, dispersion, flushing due to wave energy, emulsification, photochemical oxidation, microbial biodegradation, and adsorption to suspended matter and deposition on to the seafloor. Under sunlight, petroleum discharged at sea is subjected to photochemical modification (Harayama et al., 1999; Boehm et al., 1982; Garcia et al., 2006; Wang et al., 2013). Microbial degradation of petroleum hydrocarbon is general a long-term weathering process and represents the primary mechanisms by which petroleum hydrocarbons are eliminated from the environment. In our work, a laboratory experiment was conducted to simulate oil weathering process by microbial degradation. Composition and distribution of saturated petroleum hydrocarbons during the biodegradation of crude oils were detected, and the conventional diagnostic ratios for oil spill identification were evaluated. *n*-alkanes, a major group in crude oil, are readily biodegraded in the marine environment (Wang et al., 2013a, 2013b). The conventional diagnostic ratios of *n*-alkane displayed obvious changes after biodegradation, especially for $Pr/n\text{-}C_{17}$ and $Ph/n\text{-}C_{18}$ with relative standard deviation (RSD) more than 118.84%. It indicated that they were generally invalid for oil source identification of the less serious spill. But, terpanes, steranes and PAHs retained their molecular compositions after moderate biodegradation. Therefore, the diagnostic ratios could be used in tracking the origin and sources of hydrocarbon pollution (Garcia et al., 2006; Wang et al., 2013a, 2013b).

Bio-Stimulation and Bio-Augmentation

Oils spills occurred in the coastal area and in the sea accentuate the need for effective bioremediation methods to treat oil spills in the sea water. The technologies applied in the soil and reactors have been known for several years and relatively feasible for bioremediation (Crawford and Rosenberg, 2012).

Petroleum hydrocarbons exist naturally in all marine environments. Therefore, there has been time for numerous diverse microorganisms to evolve the capability to utilize the hydrocarbons (Atlas and Hazen, 2011). Hydrocarbon oxidizers are located in virtually all natural environments, including soil and sea water. When there is a crude oil spill, the microorganisms with the capability of degrading hydrocarbons proliferate around the spill. The local community of microorganisms will be adapted to the background. It takes a certain amount of time for their populations to increase in response to the spill due to the limitation of nutrients and oxygen, or the low bioavailability of hydrocarbons in the sea water. The spill can temporarily increase the capacity of the local oil-degrading microorganisms. However, the process may not be fast enough to prevent and restore the ecological damage. Even though oil-degrading microorganisms are found everywhere, their mere presence does not mean that environmental conditions are ideal for oil biodegradation. Environmental conditions, as well as the location, duration, and the form of an oil spill strongly affect how quickly biodegradation of microorganisms will occur (Das and Chandran, 2011; Hazen et al., 2010). There are two ways to enhance the bioremediation: (1) improve the physical interactions between bacteria and hydrocarbons by applying appropriate amount of bioemulsifiers (biosurfactant), (2) supply adequate nutrient sources to stimulate the proliferation of hydrocarbon-degrading bacteria at the polluted sites.

Addition of artificial chemical emulsifiers could rapidly disperse the crude oil and increase the solubility of the hydrocarbons. The emulsifiers performed in the oil spilled sites can have negative effects. More than 700,000 gallons of the dispersant Corexit were applied directly to the wellhead during the Deepwater Horizon blowout. Two years after the spill, studies showed that Corexit increased the toxicity of the oil by 52 times (Rico-Martinez et al., 2013). The artificially biosurfactant was also performed during the Penglai 19-3 oil spills to quickly eliminate the large oil slick on the water surface. The ecological effects of this action are still unknown.

Hydrocarbon, as the name implies, contains only hydrogen and carbon. Therefore, the bacteria require utilizable adequate nutrient sources, including nitrogen, phosphorus, sulphur, iron, oxygen and mineral elements for their proliferation (Ron and Rosenberg, 2014). The rate-limiting process in bioremediation of crude oil spill in the oligotrophic sea water is the deficiency of nitrogen and phosphorus, which can be met with the addition of fertilizers (Rosenberg et al., 1992 and 1998).

The most famous case that the nutrient fertilizer was applied is the bioremediation of the Exxon Valdez oil spill. On March 24, 1989, the oil tanker Exxon Valdez ran aground on Bligh Reef in Prince William Sound, AK. and spilled an estimated 11 million gallons of crude oil that spread as a surface slick (Wolfe et al., 1994). Much of the oil washed onto the shorelines of islands in the path of the oil, which made shoreline cleanup the primary concern. Two types of fertilizers were selected for the bioremediation. The slow release fertilizer, Customblem™ 28-8-0 (Sierra Chemicals of California), was spread at a rate of 27.8 g/m^2. The oleophilic fertilizer, Inipol™ EPA22 (Elf Aquitaine of France), was then applied at a rate of 300 g/m^2. With the first days to weeks after the spill, about 25-30% of the total hydrocarbon in the oily shorelines had been eliminated. Totally 48600 kg of N in the fertilizers were applied to the shorelines from 1989-1991 (Atlas and Hazen, 2011). Most of the crude oil had been removed from shorelines after the bioremediation process. This is the largest bioremediation of crude oil ever undertaken.

On 2 January 1997, a Russian tanker, the Nakhodka, broke apart and sank in the Sea of Japan. The incident released approximately 5000 tons of heavy oil, which was carried by winds to the coast of Japan. The local fishery cooperative association scattered approximately 250 kg of the bioremediation agent, TerraZyme™ (Oppenheimer Biotechnology) on the oil polluted shoreline rocks and concrete in the small fishing harbour of the small town in Kasumi-cho, Kinosaki-gun, Hyogo prefecture. The results showed that TerraZyme™ was able to significantly enhance the biodegradation of oil in the field. The degradation rate of the Nakhodka heavy oil is almost equivalent to the results of the laboratory tests using TerraZyme™ (Hozumi et al., 2000; Tsutsumi et al., 2000).

THE IMMOBILIZATION TECHNIQUES

There is the fact that the use of laboratory-selected bacterial strains, with high metabolic rates and broad substrate range failed in natural environments. They must be able to overcome biotic and abiotic stresses in the environment, in which they are introduced (Alexander, 1999). Macnaughton et al., (1999) demonstrated the early disappearance of the components of a microbial consortium introduced in a natural polluted site. The microbial cells could provide valuable information in lab experimentation, but were not suited to industrial applications. The study showed that the stones in the oil lake sites of the Kuwaiti desert appeared clean, which indicated that the indigenous hydrocarbon-degrading microorganisms could colonize themselves on the stones to keep them more active (Obuekwe and Al-Zarban, 1998). Therefore, the use of inoculant formulations immobilized on carrier material, particularly in packed or fluidized bed reactors could minimizethe disadvantages of the environment conditions. Carrier materials should generally provide protective niche to immobilized microorganisms, either physically, via the provision of a protective surface, or nutritionally, via the provision of a specific substrate (Gentili et al., 2006). The carriers should be nontoxic, non-pollutant, and have defined structure, maintain constant quality. There are two categories of hydrogel material for cell immobilization, natural and synthetic. For the natural hydrogel material, such as agar, agarose, polyacrylamides, κ-carrageenan, chitosan, alginate, and so on, they have low mechanical strength and durability. In contrast, strength is one of the main advantages of synthetic materials (Webb and Dervaks, 1996; Cunningham et al., 2004). The synthetic gels include poly(carbamoyl)sulphonate (PCS), poly (hydroxyethyl-methacrylate) (polyHEMA), polyacrylamide (PAM) and polyvinyl alcohol (PVA) (Jen et al., 1996).

Immobilized cells have been used and studied for the bioremediation of numerous toxic chemicals. A laboratory-scale study showed that the chitin and chitosan flakes obtained from shrimp wastes are potential good carrier materials for the hydrocarbon-degrading bacterial strain. The immobilized microbial strain could remove 60% of hydrocarbons compared with the controls. While, the strain was inoculated without carrier removed only 30% of the hydrocarbons in the sea water microcosm test (Gentili et al., 2006). Immobilization not only simplifies separation and recovery of the immobilized bacteria and the binding agent, but it also makes the application reusable, which reduces the overall cost (Wang et al., 2012).

Sawdust, Styrofoam and wheat bran were used as carriers for the immobilization of two hydrocarbon-degrading bacterial strains, *Arthrobacter* sp. and *Bacillus* sp. (Obuekwe and Al-Muttawa, 2001). Under the low nutrient conditions, stable exopolysaccharide-mediated immobilized cultures were formed. The immobilized cultures could survive and still retain the capability of degrading hydrocarbons after air-dried at room temperature. Radwan and Al-Hasan (2001) reported that hydrocarbon-degrading microbial consortia in biofilms naturally coated gravel particles in the intertidal zone of Arabian Gulf coast and could attenuate hydrocarbons in contaminated sea water. Further, the same group developed a method to artificially establish biofilms including bacterial strains, picoplankton and diatoms, which showed an attenuation effect on crude oil contaminated sea water samples (Al-Awadhi et al., 2003).

Furthermore, immobilized materials have comparatively longer operating lifetimes due to an enhanced stability of the macromolecules or cells and, consequentially, protection from adverse conditions. Biomass immobilized in a range of inert materials like silica, polyacrlyamide, polymethane and polysulfone, has been used in a variety of bioreactor. Wilson and Bradley (1996) immobilized *Pseudomonas fluorescens* on a commercially available biosupport (Biofix) and a biosorbent (Drizit). Then, they used the free suspension and the immobilized bacteria as bioremediation agents in an aqueous system with petrol as the carbon source. The study indicated that the immobilization resulted in a combination of increased contact between cell and hydrocarbon droplets and enhanced levels of rhamnolipids production.

Wang et al. (2012) compared the efficiencies of free and immobilized bacterial cultures of petroleum hydrocarbon degrading bacteria. The microbial communities with high tolerance and high degrading capability of crude oil were obtained and then immobilized in sodium alginate beads and sodium alginate-diatomite beads. The immobilized cells in both beads resulted in higher removal of crude oil than free cells at the same inoculation concentrations.

Moreover, the sodium alginate-diatomite beads demonstrated advantages over the sodium alginate beads.

Cunningham et al., (2004) used polyvinyl alcohol (PVA) cryogelation as an entrapment matrix and indigenous microorganisms at the site. The laboratory biopileswere constructed to compare immobilized bio-augmentation with liquid culture bio-augmentation and bio-stimulation. The immobilized systems were found to be the most successful in terms of percentage removal of diesel after 32 days. The co-immobilization system containing polyvinyl alcohol-entrapped microorganisms showed the greatest removal rate of diesel.

High salinities or wide salinity ranges make the oil pollution difficult to be remediated by the conventional methods, especially in the estuaries, beaches, salt marshes, sea water. Few studies were conducted to deal with the effect of salinity on microbial degradation of crude oil, although the detrimental effect on microbial activities has been reported widely. Díaz et al., (2002) reported firstly on the immobilization of an extremely halotolerant bacterial population for bioremediation of hydrocarbons over a wide range of salinities. They found that the bacterial consortium MPD-M, which was isolated from the sediment associated with Colombian mangrove roots, was effective in the treatment of hydrocarbons in water with salinities varying from 0 to 180 g/L. The biodegradation efficiency increased when the bacterial cells were immobilized on polypropylene fibers. Compared to the free cells, the immobilized cells demonstrated higher biodegradation rates (4 and 7 times), even at the highest salinity (180 g/L). Their study indicated that the increased stability of the intracellular activities could be attributed to the protective effects of the biofilm against physicochemical stresses. The use of oleophilic carrier also could increase the availability of the crude oil to the bacterial consortium.

After the Penglai 19-3 oil spill in Bohai Bay, China, we isolated two efficient hydrocarbon degraders from the contaminated sediments. 16S rRNA gene sequencing and phylogenetic analysis indicated that the two bacterial strains affiliated with the genera *Vibrio* and *Acinetobacter*. Subsequently, the bacterial cells were immobilized on the surface of cotton fibers. Cotton fibers were used as crude oil sorbent as well as carriers for bacterial immobilization. We interestingly noticed that one of the marine bacteria strain, *Acinetobacter* sp. could strongly bind to the cotton fibers (Figure 1b). *Vibrio* sp. strain showed no self-adherent to the cotton fibers (Figure 1a). The coherent mass of *Acinetobacter* sp. grown on the carrier surface, and the inter-connection between individual bacterial cells, suggested that these bacterial strains were likely to produce extracellular polysaccharides to enhance the immobilization. Hence, both immobilized *Acinetobacter* sp. strains and the planktonic bacterial cells were chosen to compare their degradation rate. Our results indicated that both planktonic and immobilized bacteria showed relatively high biodegradation rates. The degradation activities of both planktonic and immobilized bacteria were not affected significantly even the NaCl concentration reached 70 g/L. The degradation rates of both planktonic and immobilized bacteria were above 60% with the pH ranged from 5.6 to 8.6. The efficiency of saturated hydrocarbons degradation by immobilized bacterial cells increased about 30% compared to the planktonic bacterial cells. The cotton fibers were proved to be a useful crude oil adsorbent and good bacterial carriers (Lin et al., 2014).

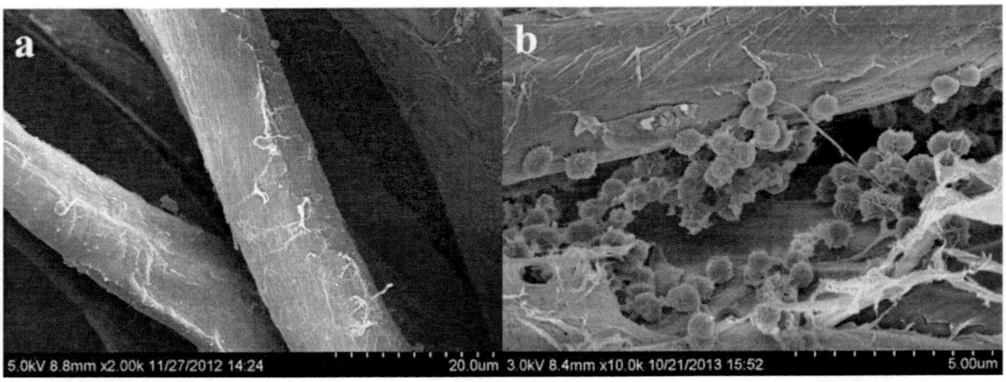

Figure 1. SEM images of the typical adhesion of the bacterial isolates *Vibrio* sp. (a) and *Acinetobacter* sp. (b) on the surface of cotton fibers after 24 h incubation.

Research Needs

Several disastrous oil spills occurred in the marine ecosystems, such as Exxon Valdez in Alaska, Deepwater Horizon oil spill in the Gulf of Mexico, and Penglai 19-3 oil spill in Bohai Bay, China, showed us that indigenous microorganisms play key roles in eliminating spilled crude oil. At the time of Exxon Valdez, advanced molecular techniques were not available to characterize or monitor the microbial communities before and after the oil spill. The growing availability of advanced molecular technologies and genomic sequencing information of microorganisms capable of degrading hydrocarbons provides a start to comprehensively understand the interactions between microorganisms and hydrocarbons. The extensive molecular analyses of microbial communities have been performed in the Gulf of Mexico following the Deepwater Horizon oil spill, which give us the opportunity to know the proliferation and succession of the microbial communities. However, to completely understand the dynamic changes of the microbial communities and the microbial network, genes, proteins, and metabolic pathways alone will not be enough. Systems biology approach integrating biochemistry, thermodynamics, meta-genomic sequencing, regulatory and metabolic network analysis, towards understanding the metabolism of microbial communities is necessary in the future study. The successful prediction of system behaviour in a given type of environment will lead us to the more efficient and environmentally-friendly bioremediation process.

References

Al-Awadhi H, Al-Hasan RH, Sorkhoh NA, Salamah S, Radwan SS. Establishing oil-degrading biofilms on gravel particles and glass plates. *International Biodeterioration & Biodegradation.* 2003. 51: 181-185.

Alexander M. Inoculation. In: *Biodegradation and Bioremediation,* 2nd, ed. Academic Press, San Diego, CA, USA, 1999. pp. 299-311.

Al-Majed AA, Adebayo AR, Hossain ME. A sustainable approach to controlling oil spills. *Journal of Environmental Management*. 2012. 113: 213-227.

April TM, Foght JM, Currah RS. Hydrocarbon degrading filamentous fungi isolated from flare pit soils in northern and western Canada. *Canadian Journal of Microbiology*. 2000. 46(1): 38-49.

Atlas RM, Hazen TC. Oil biodegradation and bioremediation: A tale of the two worst spills in U.S history. *Environmental Science & Technology*. 2011. 45: 6709-6715.

Boehm PD, Fiest DL, Mackay D, Paterson S. Physical chemical weathering of petroleum hydrocarbons from the Ixtoc I blowout: chemical measurements and a weathering model. *Environmental Science & Technology*. 1982. 16: 498-505.

Crawford RL, Rosenberg E. Bioremediation. A recent review on the technology of bioremediation on land and in closed reactors. *In the prokaryotes,* 4th ed., edited by Rosenberg E, Delong E, Thompson F, Lorey S, Stackebrandt E. new York; Springer; 2012.

Crone TJ, Tolstoy M. Magnitude of the 2010 Gulf of Mexico oil leak. *Science*. 2010. 330: 634-634.

Cunningham CJ, Ivshina IB, Lozinsky VI, Kuyukina MS, Philp JC. Bioremediation of diesel-contaminated soil by microorganisms immobilized in polyvinyl alcohol. *International Biodeterioration & Biodegradation*. 2004. 54: 167-174.

Das N, Chandran P. Microbial Degradation of Petroleum Hydrocarbon Contaminants: An Overview. *Biotechnology Research International*. Volume 2011, Article ID 941810, 13 pages.

Díaz MP, Boyd KG, Grigson SJW, Burgess JG. Biodegradation of crude oil across a wide range of salinities by an extremely halotolerant bacterial consortium MPD-M, immobilized onto polypropylene fibers. *Biotechnology and Bioengineering*. 2002. 79: 145-153.

Garcia de O, Grimalt JO. GC and GC-MS characterization of crude oil transformation in sediments and microbial mat samples after the 1991 oil spill in the Saudi Arabian Gulf coast. *Environmental Pollution*. 2006. 139: 523-531.

Gentili AR, Cubitto MA, Ferrero M, Rodriguéz. Bioremediation of crude oil polluted seawater by a hydrocarbon-degrading bacterial strain immobilized on chitin and chitosan flakes. *International Biodeterioration and Biodegradation*. 2006. 57(4): 222-228.

Harayama S, Kishira H, Kasai Y, Shutsubo K. Petroleum biodegradation in marine environments. *Journal of Molecular Microbiology and Biotechnology*. 1999. 1: 63-70.

Hazen TC, Dubinsky EA, DeSantis TZ, Andersen GL, Piceno YM, Singh N, Jansson JK, Probst A, Borglin SE, Fortney JL, Stringfellow WT, Bill M, Conrad ME, Tom LM, Chavarria KL, Alusi TR, Lamendella R, Joyner DC, Spier C, Baelum J, Auer M, Zemla ML, Chakraborty R, Sonnenthal EL, D'haeseleer P, Holman H-YN, Osman S, Lu ZM, Van Nostrand JD, Deng Y, Zhou JZ, Mason OU. Deep-sea oil plume enriches indigenous oil-degrading bacteria. *Science*. 2010. 330: 204-208.

Hozumi T, Tsutsumi H, Kono M. Bioremediation on the shore after an oil spill from the Nakhodka in the Sea of Japan. I. Chemistry and characteristics of the heavy oil loaded on the Nakhodka and biodegradation tests on the oil by a bioremediation agent with microbiological cultures in the laboratory. *Marine Pollution Bulletin*. 2000. 40: 308-314.

Jen AC, Wake MC, Mikos AG. Review: hydrogels for cell immobilization. *Biotechnology and Bioengineering*. 1996. 50: 254-272.

Joye SB, MacDonald IR, Leifer I, Asper V. Magnitude and oxidation potential of hydrocarbon gases released from the BP oil well blowout. *Nature Geoscience.* 2011. 4: 160-164.

Kessler JD, Valentine DL, Redmond MC, Du MR, Chan EW, Mendes SD, Quiroz EW, Villanueva CJ, Shusta SS, Werra LM, Yvon-lewis SA, Weber TC. A persistent oxygen anomaly reveals the fate of spilled methane in the Deep Gulf of Mexico. *Science.* 2011. 331: 312-315.

Kostka JE, Prakash M, Overholt WA, Green SJ, Freyer G, Canion A, Delgardio J, Norton N, Hazen TC, Huettel M. Hydrocarbon-degrading bacteria and the bacterial community response in Gulf of Mexico beaches and impacted by the Deep water Horizon oil spill. *Applied Environmental Microbiology.* 2011. 77:7962-7974.

Leahy JG, Colwell RR. Microbial degradation of hydrocarbons in the environment, *Microbiological Reviews.* 1990. 54(3): 305-315.

Lin M, Liu YH, Chen WW, Wang H, Hu XK. Use of bacteria-immobilized cotton fibers to absorb and degrade crude oil. *International Biodeterioration & Biodegradation.* 2014. 88: 8-12.

MacDonald IR, Leifer I, Sassen R, Stine P, Mitchell R, GuinassoJr N. Transfer of hydrocarbons from natural seeps to the water column and atmosphere. *Geofluids.* 2002. 2(2): 95-107.

Macnaughton SJ, Stephen JR, Vesosa AD, Davis GA, Chang YJ, White DC. Microbial population changes during bioremediation of an experimental oil spill. *Applied Environmental Microbiology.* 1999. 65: 3566-3574.

Martins dos Santos VAP, Yakimov MM, Timmis KN, Golyshin PN. Genomic insights into oil biodegradation in marine systems. *Microbial Biodegradation: Genomics and Molecular Biology.* 2008. Caister Academic Press.

Medina-Bellver JI, Marin P, Delgado A, Rodríguez-Sánchez A, Reyes E, Ramos JL, Marqués S. Evidence for in situ crude oil biodegradation after the Prestige oil spill. *Environmental Microbiology.* 2005. 7(6): 773-779.

Obuekwe CO, Al-Muttawa EM. Self-immobilized bacterial cultures with potential for application as ready-to-use seeds for petroleum bioremediation. *Biotechnology Letters.* 2001. 23: 1025-1032.

Obuekwe CO, Al-Zarban SS. Bioremediation of crude oil pollution in the Kuwaiti desert: the role of adherent microorganisms. *Environment International.* 1998. 24: 823-834.

Radwan SS, Al-Hasan, RH. Potential application of coastal biofilm-coated gravel particles for treating oily waste. *Aquatic Microbial Ecology.* 2001. 23: 113-117.

Rico-Martinez R, Snell TW, Shearer TL. Synergistic toxicity of Macondo crude oil and dispersant Corexit 9500A to the *Brachionusplicatilis* species complex (Rotifera). *Environmental Pollution.* 2013. 173: 5-10.

Ron EZ, Rosenberg E. Enhanced bioremediation of oil spills in the sea. *Current Opinion in Biotechnology.* 2014. 27: 191-194.

Rosenberg E, Legmann R, Kushmaro A, Taube R, Adler E, Ron E. Petroleum bioremediation - a multiphase problem. *Biodegradation.* 1992. 3: 337-350.

Rosenberg E, Navon-Venezia S, Zilber-Rosenberg I, Ron EZ. Rate-limiting steps in the biodegradation of hydrocarbons. In soil and aquifer pollution. Edited by Rubin H, Narkis N., *Canberry J.* 1998: 159-172.

Tsutsumi H, Kono M, Takai K, Manabe T, Haraguchi M, Yamanoto I, Oppenheimer C. Bioremediation on the shore after an oil spill from the Nakhodka in the Sea of Japan. III. Field tests of a bioremediation agent with microbiological cultures for the treatment of an oil spill. *Marine Pollution Bulletin.* 2000. 40: 320-324.

Wang CY Chen B, Zhang BY, He SJ, Zhao MM. Fingerprint and weathering characteristics of crude oils after Dalian oil spill, China. *Marine Pollution Bulletin.* 2013a. 71: 64-68.

Wang CY, Hu XK, He SJ, Liu X, Zhao MM. Source diagnostic and weathering indicators of oil spills utilizing bicyclic sesquiterpanes. *Acta Oceanologica Sinica. 2013b.* 32(8): 79-84.

Wang H, Wang CX, Lin M, Sun XN, Wang CY, Hu XK. Phylogenetic diversity of bacterial communities associated with bioremediation of crude oil in microcosms. *International Biodeterioration and Biodegradation.* 2013. 85: 400-406.

Wang ZY, XU Y, Wang HY, Zhao J, Gao DM, Li FM, Xing B. Biodegradation of crude oil in contaminated soils by free and immobilized microorganisms. *Pedosphere.* 2012. 22: 717-725.

Webb C, Dervakos GA. *Studies in Viable Cell Immobilization.* 1996. Academic Press, London.

Wilson NG, Bradley G. The effect of immobilization on rhamnolipid production by *Pseudomonas fluorescens. Journal of Applied Bacteriology.* 1996. 81: 525-530.

Wolfe DA, Hameedi MJ, Galt JA, Watabayashi D, Short J, O'Clair C, Rice S, Michel J, Payne JR, Braddock J, Hanna S, Sale D. Fate of the oil spilled from the T/V Exxon Valdez in Prince William Sound, Alaska. *Environmental Science & Technology.* 1994. 28: 561A-568A.

Yakimov MM, Timmis KN, Golyshin PN. Obligate oil-degrading marine bacteria. *Current Opinionin Biotechnology.* 2007. 18(3): 257-66.

In: Oil Spills
Editor: Adam Clifton

ISBN: 978-1-63321-515-3

Chapter 9

A Comparison of Evaluation Methods for Decision-Making of Oil Spill Response Strategy

Xin Liu[1,]**, *Jie Guo[1]*, *Xiaoke Hu[1]*, *Chuanyuan Wang[1]*, *Cheng Tang[1]* and *Guangchen Pan[2]

[1]Key Laboratory of Coastal Zone Environmental Processes, CAS; Shandong Provincial Key Laboratory of Coastal Zone Environmental Processes;Yantai Institute of Coastal Zone Research, Chinese Academy of Sciences, Yantai, P. R. China

[2]The University of Yantai, Yantai, P. R.China

Abstract

Oil spill often leads to significant, long-term impacts on the environment. This requires a best decision to be made at the right moment. Hence, evaluating combat options in a sufficient way is still an ongoing concern challenging decision makers. A set of analysis tools including multi-criteria analysis (MCA), fuzzy comprehensive evaluation (FCE) and monetary evaluation model (MEM) have been developed and applied recently in the management for oil spill contingency. Following a close comparison of the methods, selections of evaluation methods are finally summarized for development of good practice in the context of integrated coastal zone management (ICZM).

Keywords: Oil spill, evaluation method, comparison

1. Introduction

One major concern of an Integrated Coastal Zone Management (ICZM) in many areas around the globe is related to the preparedness against accidental oil spills. Decision making

*Corresponding author. Email: xin.icbm@gmail.com. Tel. 0086(0)5352109195; Fax. 0086(0)5352109000

in response to immanent or potential oil pollution often encompasses multiple and conflicting objectives and has to face various groups of uncertainties, constituting a challenge for operational contingency management. Ideally, multiple objectives can be aggregated into a single objective function, which allows for identifying one best solution by optimization. However, this solution strongly depends on the aggregation scheme. Here three decision making models adapted to the multi-objective optimization problem are developed in the presence of parametric uncertainties: (1) the multi-criteria analysis (MCA), (2) fuzzy comprehensive evaluation (FCE) and (3) the monetary evaluation model (MEM). MCA is a straightforward method to aggregate a variety of ecological and socio-economic criteria, which are measured in different metrics (Almasri and Kaluarachchi, 2005; Volchko et al., 2014; Fontana et al., 2013). Since the FCE is capable to process lexical data, it provides a clear and traceable structure to integrate different stakeholders into the decision making process (Wang et al., 2012; Zuo et al., 2009; Cheng and Qian, 2010). A FCE further aids to determine a critical criterion by sensitivity analysis, which guides decision makers during the set-up of preferences. A MEM transfers impact intensities in various sectors to monetary values as these may more intuitively illustrate to both the public and professionals the effects of management and planning in both short and long terms (Subade and Francisco, 2014; Can and Alp, 2012; Tagliafierro et al., 2013). The economic approach helps to select a cost-effective pollution reduction measures in the particular case of oil spills. These evaluation models mentioned above have been applied to solve a variety of marine environmental problems (Börger et al., 2014; Laurans et al., 2013; Gao and Hailu, 2012; Wu et al., 2012; Tammi and Kalliola, 2014; Tuda et al., 2014) To optimize combat options, Liu and Wirtz (2006a, 2006b), Wirtz and Liu (2006) and Liu et al. (2009) have developed and tested these methods in oil spill cases to aid decision making for contingency management.

2. Overview of Selected Evaluation Methods

2.1. Multi-criteria Analysis (MCA)

Multi-criteria analysis presumes that respondents have a preference function to assess performance of alternatives based on selected criteria (In et al., 2001). The most common form of a preference function is:

$$\hat{S}_j = \sum_{i=1}^{n} W_i * S_{ij} \tag{1}$$

Where S_{ij} is the relative score of alternative j on criterion i and W_i indicates the relative importance of criterion i. $\hat{S}_j$ is the overall score for alternative j. The final outcome is a preferred option based on rankings of the overall performances of options.

2.2. Fuzzy Comprehensive Evaluation (FCE)

FCE is characterized by dealing with lexically imprecise and incomplete knowledge. Following Ji et al. (2000) and Liu and Wirtz (2006b), it has three steps for application. Firstly, the first-order FCE set for alternative j with respect to the criterion i is an outcome of multiplying the pre-defined first order fuzzy degree assignment matrix (R_i) and the fuzzy degree vector (μ_{ij}) as follows:

$$B_{ij} = \mu_{ij} \circ R_i \quad (2)$$

where B_{ij} represents a fuzzy performance of alternative j with respect to the ith criterion. Equally spaced damage levels ranging from 0 to 1 are predefined for the damage degree: $\varsigma = \{\varsigma_1, \varsigma_2, \cdots, \varsigma_{11}\} = \{0, 0.1, \cdots, 0.9, 1.0\}$, in which 0 represents no damage, while 1.0 denotes a completely destroyed coastal area. For any criterion i, a first order fuzzy degree assignment matrix (R_i) is constructed, in which each element represents a fuzzy degree of each kind of damage grade against each of these 11 coastal damage levels. For construction of R_i and calculation of μ_{ij} , Liu and Wirtz (2006b) can be referred. Respondents may express their opinions about the importance of the selected criteria in a qualitative way such as important, relevant and unimportant. Numerical weights are resulted from transferring these linguistic terms. The second order FCE is then obtained by multiplying the numerical weight values with the first order fuzzy performance of the specific alternative as follows:

$$K_j = W \circ B_j = \{k_j^1, k_j^2, \cdots, k_j^{11}\} \quad (3)$$

Finally, the overall coastal impact performance(CIP) of alternative j can be determined by combining equation K_j and ς as follows:

$$CIP_j = \sum_{p=1}^{11} k_p \varsigma_p \Big/ \sum_{p=1}^{11} k_p \quad (4)$$

2.3. Monetary Evaluation Model (MEM)

Monetary evaluation model is referring to calculation of damaged environmental resources in money terms. The lost value (V) of a specific habitat polluted by oil spills is calculated as:

$$V = \sum_{i=1}^{n} M_i * Q_i * LS_i \quad (5)$$

Where M is the monetary value of the natural resource i per unit per year, wich can be measured by a variety of environmental valuation methods; Q is the total quantities of the injured natural resource i and LS is its lost service-years. When the injury is primarily to economic sectors such as the fishery and tourism, the above formulation can be rewritten according to Liu and Wirtz (2006a),

$$IL = \sum_{i=1}^{n} YI_i * (\frac{A_i}{TA_i}) LS_i \tag{6}$$

for a specific economic sector i, IL is its income losses, YI indicates its yearly income, A and TA are the polluted area and the total area of the economic sector i in the affected region, respectively. LS is its lost service as mentioned above.

3. Comparison of Methods

3.1. Procedures

At first glance, the three approaches seem quite different. On closer examination, their procedures appear similar. This can be seen more clearly in Figure 1, which depicts a clear and traceable decision making process for the three different methods. The six steps and their order are common to both approaches of MCA and FCE. While MEM only share five out of the six main steps, as the weights between the criteria are not required for carrying out an appraisal by MEM. At this level of description, these methods are very similar. However, examination of the features unique to each shows considerable difference. These differences lie on the steps of overall evaluation and analysis which are marked with a star in Figure 1.

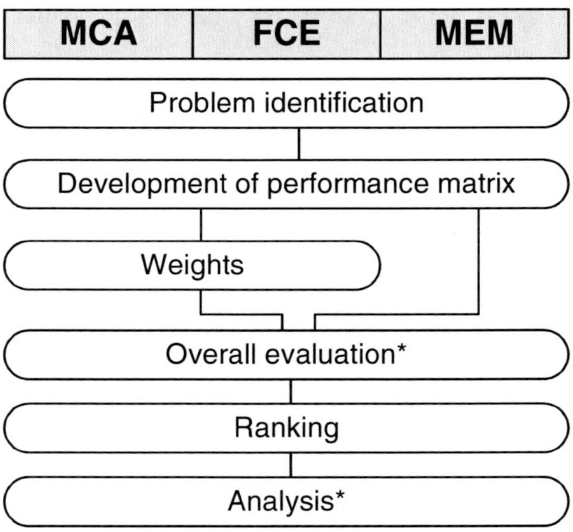

Figure 1. Similarities among these methods.

Table 1. Unique features to each method

	Methods		
	MCA	FCE	MEM
Involvement of stakeholders	yes	yes	no
Inclusion of long-term effects	no	no	yes
Capability of dealing with data in multiple dimensions	strong	Very strong	Sort of strong
Requirement of data	Quantitative data	No requirement	Quantitative data in the specific dimension
Intuitive indicator	Not intuitive	Sort of intuitive	Very intuitive
Internal set-up procedure	Very simple	Complex	Simple
Uncertainty introduced	1.Normalization function 2.Linear aggregation model	1.Membership function 2.First-order assignment matrix	1.Unit price defined 2.Recovery function
Potential analysis	1. Effect of changing weight on the ranking 2. Critical criteria determination		1.The benefit cost analysis 2.The cost effectiveness analysis

3.2. Unique Features

Table 1 summarizes these differences in a great detail. An involvement of stakeholders is the feature unique to MCA and FCE. MEM could provide a long-term effect of the combat strategy responding to the incident by discounting of all its costs and benefits to present values. FCE has no requirement for data, while quantitative data are preferred in MCA and MEM only deal with quantitative data in specific dimensions. Thus, in terms of dealing with data in multiple dimensions, both MCA and FCE show a stronger capability than MEM. The indicator used in MCA is score which is not intuitive to describe the effects of oil pollution. In FCE we use the a pollution index as the indicator, which varies with the internal of [0,1]. In MEM money as the indicator intuitively provides both the public and decision-makers the effects of the pollution. By a comparison of their internal set-ups, MCA and MEM are easier to understand and use than FCE. Moreover, a variety of uncertainties may be introduced during the process of internal set-ups. For example, normalization function and linear aggregation are two potential sources to introduce uncertainties in MCA. Meanwhile, unnecessary uncertainties could be resulted from unsuitable specifications of membership function and first-order assignment matrix in FCE. When MEM is applied, man should be careful to check the effect of setups of unit price of environmental goods and their recovery function on the ranking result of the alternatives. Finally, both of MCA and FCE provide the same analyses such as the effects of weights on the ranking orders and the determination of critical criteria. While, MEM provides the benefit cost analysis and the cost effectiveness analysis.

Table 2. Advantages and disadvantages for each method

Methods	Advantages	Disadvantages
MCA	• Easiness to understand and use; • Involvement of stakeholders; • The benchmark, if all data are measured in one dimension; • Enable to determine the critical criteria	• It may not be adequate to deal with data whose relation are non-linear • Quantitative data needed
FCE	• Capable of dealing with fuzzy data • Involvement of stakeholders; • Enable to determine the critical criteria • Rational approach to integrate data	• Complex procedure of internal setups • Introduction of uncertainty due to a lack of expert knowledge
MEM	• Very rational approach to integrate data • Provide intuitively pollution effects in money terms • Incorporate long-term effects of incident • Possible to determine a worthwhile investment of combat facilities	• Lack of involvement of stakeholders • Need expert knowledge • Specific requirements of data • Quantitative data needed

3.3. Advantages and Disadvantages

Each method has its own advantages and disadvantages. For each method, the main benefits and costs are summarized in Table 2. For example, advantages of MCA include that this method is easy to be applied. It allows an involvement of stakeholders. The result derived from this method will provide a benchmark, if all data are measured in one dimension. Also it enables to determine the critical criteria. And its disadvantages cover the requirements for the data and limited capability to deal with the data whose relation are non-linear. Given that data are not linearly related with each other, both FCE and MEM can be regarded as more rational methods than MCA to integrated data. Such an aspect has been listed in Table 2. as one of the benefits in FCE and MEM respectively.

4. Selection of Methods

The selection of these methods highly depends on the obtained data and knowledge, decision process and the desired result. Table 3 suggests which method should be preferred corresponding to a variety of situations. In some particular situations, more than one method are suggested.

Table 3. Selection of evaluation method

Where		Methods preferred
Decision process	Bottom-up decision	MCA/FCE
	Top-down decision	MCA/FCE/MEM
	Making a group decision	MCA/FCE
Data and knowledge available	Relation between data exhibits non-linearity	FCE/MEM
	Lexical data	FCE
	Single dimension data	MCA
	Multiple dimension data	FCE/ MCA /MEM
	No expert knowledge	MCA/FCE
Desired result	Consideration of long-term effects	MEM
	Determine the critical criteria	FCE/MCA
	Determine a worthwhile investment	MEM
	An intuitively comparable indicator of pollution effects wanted	MEM

CONCLUSION

In terms of appraising oil spill combat options, the paper reviewed and compared three different evaluation methods: Multi-critieria analysis (MCA), Fuzzy comprehensive evaluation (FCE) and Monetary evaluation model (MEM). Based on their application procedures, unique features, advantages and disadvantages, selection of evaluation methods are further summarized to any specific situation. It is important to recognize that both MCA and FCE were not developed from a base of economics. Hence, it has not developed procedures for valuating oil spill combat options' benefits and costs that are specially geared to the public interests.

ACKNOWLEDGMENTS

This work is jointly supported by a number of funds numbered with kzcx2-yw-qn208 from CAS, 41371483 from the Natural Science Foundation, China and KZZD-EW-14 from the Key Research Program of the Chinese Academy of Sciences.

REFERENCES

Almasri, M. N., and Kaluarachchi, J. J., 2005. Multi-criteria decision analysis for the optimal management of nitrate contamination of aquifers. *Journal of Environmental Management 74,* 365–381.

Börger,T., Beaumont,N. J., Pendleton, L., Boyle,K. J., Cooper, P., Fletcher,S., Haab,T., Hanemann,M., Hooper,T. L., Hussain, S.S., Portela, R., Stithou,M., Stockill,J., Taylor,T.,

Austen, M.C.,2014. Incorporating ecosystem services in marine planning: The role of valuation. *Marine Policy, 46,* 161-170.

Can, Ö and Alp, E., 2012. Valuation of environmental improvements in a specially protected marine area: A choice experiment approach in Göcek Bay, Turkey. Science of The Total Environment,439, 291-298.

Cheng,C. Y. and Qian, X., 2010. Evaluation of Emergency Planning for Water Pollution Incidents in Reservoir Based on Fuzzy Comprehensive Assessment. *Procedia Environmental Sciences, 2,* 566-570.

Fontana,V., Radtke,A., Fedrigotti,V. B., Tappeiner,U., Tasser,E., Zerbe, S., Buchholz, T.,2013. Comparing land-use alternatives: Using the ecosystem services concept to define a multi-criteria decision analysis. *Ecological Economics, 93,* 128-136.

Gao,L., Hailu, A., 2012.Ranking management strategies with complex outcomes: An AHP-fuzzy evaluation of recreational fishing using an integrated agent-based model of a coral reef ecosystem.*Environmental Modelling & Software*, 31, 3-18.

In, H., Olson, D., Rodgers, T., 2001. A requirements negotiation model based on multi-criteria analysis. *In the Proceedings of the Fifth International Symposium on Requirements Engineering (RE'01),* pages 312-313.

Ji, S., Li, X., Ma, Y., 2000. Optimal tolerance allocation based on fuzzy comprehensive evaluation and genetic algorithm. *International Journal of Advanced Manufacturing Technology 16,* 451-468.

Laurans,Y., Pascal,N., Binet,T.,Brander,L., Clua,E., David, G., Rojat,D., Seidl, A., 2013. Economic valuation of ecosystem services from coral reefs in the South Pacific: Taking stock of recent experience. *Journal of Environmental Management, 116,* 135-144.

Liu, X., Wirtz, K.W., 2006a. Total oil spill costs and compensations. *Maritime Policy and Management, 33,* 49-60.

Liu, X., Wirtz, K.W., 2006b. Consensus oriented, fuzzified decision support for oil for spill contingency management. *Journal of Hazardous Materials, 134*, 27-35.

Liu, X., Wirtz, K. W.,Kannen, A., Kraft, D., 2009. Willingness to pay among households to prevent coastal resources from polluting by oil spills: A pilot survey. *Marine Pollution Bulletin, 58,* 1514-1521.

Subade,R. F. And bFrancisco, H., A.,2014. Do non-users value coral reefs?: Economic valuation of conserving Tubbataha Reefs, Philippines. *Ecological Economics, 102,* 24-32.

Tagliafierro,C., Longo, A., Eetvelde, V.V., Antrop, M., Hutchinson, W. G.,2013. Landscape economic valuation by integrating landscape ecology into landscape economics. *Environmental Science & Policy, 32,* 26-36.

Tammi,I., Kalliola, R., 2014. Spatial MCDA in marine planning: Experiences from the Mediterranean and Baltic Seas. *Marine Policy, 48,* 73-83.

Tuda, A. O., Stevens, T.F., Rodwel, L.D.,2014. Resolving coastal conflicts using marine spatial planning. *Journal of Environmental Management, 133,* 59-68.

Volchko,Y., Norrman,J., Rosén, L., Bergknut,M., Josefsson, S., Söderqvist,T., Norberg, T., Wiberg,K., Tysklind,M., 2014. Using soil function evaluation in multi-criteria decision analysis for sustainability appraisal of remediation alternatives. *Science of The Total Environment, 485–486,* 785-791.

Wang,Y., Yang,W., Li,M., Liu, X.,2012. Risk assessment of floor water inrush in coal mines based on secondary fuzzy comprehensive evaluation. *International Journal of Rock Mechanics and Mining Sciences, 52,* 50-55.

Zuo, R., Cheng, Q., Frederik, P., Agterberg, F.P., 2009. Application of a hybrid method combining multilevel fuzzy comprehensive evaluation with asymmetric fuzzy relation analysis to mapping prospectivity. *Ore Geology Reviews, 35,* 101-108.

Wirtz, K. W. and Liu, x., 2006. Integrating economy, ecology and uncertainty in an oil-spill DSS: The Prestige accident in Spain, 2002. *Estuarine, Coastal and Shelf Sciences, 70,* 525-532.

Wu, H.Y., Chen, K.L., Chen,Z.H., Chen,Q.H., Qiu, Y.P., Wu, J.C., J.F., Zhang, J.F., 2012. Evaluation for the ecological quality status of coastal waters in East China Sea using fuzzy integrated assessment method. *Marine Pollution Bulletin, 64,* 546-555.

In: Oil Spills
Editor: Adam Clifton

ISBN: 978-1-63321-515-3

Chapter 10

SPECIES SPECIFIC RESPONSES OF COMMON FRESHWATER WETLAND PHYTOPLANKTON TO CRUDE OIL BY APPLICATION OF BIOASSAYS

***Paul Oberholster*[1,2]*, Paul Cheng*[1]*, Christoff Truter*[3]*, Liesl Hill*[4] *and Anna-Maria Botha*[5]**

[1]CSIR Natural Resources and the Environment, Stellenbosch, South Africa
[2]Department of Paraclinical Sciences, Faculty of Veterinary Science, University of Pretoria, Onderstepoort, South Africa
[3]Department of Botany and Zoology, University of Stellenbosch, Matieland, Stellenbosch, South Africa
[4]CSIR Natural Resources and the Environment, Pretoria, South Africa
[5]Department of Genetics, University of Stellenbosch, Matieland, Stellenbosch, South Africa

ABSTRACT

Phytoplankton are the major primary producers in freshwater environments and play a crucial role in the inflow of primary energy to freshwater food webs. The presence of phytoplankton biomass in freshwater systems can reflect changes in water quality and can act as a sensitive indicator of biological integrity and physico-chemical condi-tions of a water system. Several studies have been conducted on oil spills in the marine environment and its subsequent toxicity to organisms. Little is however known about the effects of crude oil on freshwater phyto-plankton species inhabiting wetlands. The objectives of the current study were to examine the effects of crude oil on selected common freshwater wetland phytoplankton species and to establish whether responses of freshwater phytoplankton to different concentrations of crude oil under laboratory conditions, were species specific. It was evident from the bio-assays conducted that the selected concentrations of crude oil (25 – 2500 $mg.l^{-1}$) had less pronounced toxic effects on the isolated wetland phyto-plankton species but instead promoted the growth of the green algal *Chlorella vulgaris* and *Chlorella protothecoides* and the cyanobacterium *Chrococcus dispersus*. In contrast, the growth of the benthic filamentous green alga

Oedogonium croasdaleae responded negatively to all the concentrations of crude oil evaluated. Sub-lethal levels of the crude oil induced a shift in the concentrations of chlorophyll *a* and *b*, as well as *β*-carotene, depending on the algal species and the concentration of crude oil.

The current study showed that the variability in the resistance to crude oil pollution of specific algal species may lead to uncharacteristic species succession with more tolerant species becoming the dominant group in depression wetlands. In a crude oil contaminated freshwater wet-land, this could potentially result in eutrophication which may in turn have direct consequences for wetland functioning.

INTRODUCTION

The response of phytoplankton populations to contaminated environments is ecologically relevant, since their responses will undoubtedly affect the upper trophic levels in an aquatic system. O'Brien and Dixon (1976) showed that the availability of petroleum related hydrocarbons influenced the response of phytoplankton species differently in terms of growth rate and photosynthetic assimilation. Crude oil, which consists of a mixture of up to 10000 different types of highly toxic hydrocarbons, interferes with the photosynthetic processes of phytoplankton by inhibiting the function of photosystem II (PSII) (Dawes, 1998). At a cellular level, phytoplankton exposed to crude oil products show reduced DNA, RNA, and protein content, while polycyclic aromatic hydrocarbon (PAHs) reduces expression of genes for photosynthetic pigments (El-Sheekh et al., 2000).

The stress of oil pollution and the natural variability in the tolerance of different phytoplankton groups can result in an abnormal species succession in water ecosystems (Jiang et al., 2010; Oberholster et al., 2010). The less oil tolerant species, which may be the dominant algal species in a water body can decrease in cell numbers or be abolished, while the more resistant algal groups may gradually become dominant (Jiang et al., 2010; Oberholster et al., 2010).

Oil tolerant groups may maximize their reproductive potential because competition for nutrients and solar radiation decreases after the elimination of the more susceptible species (O'Brien and Dixon 1976). Previous studies have shown that the diversity of phytoplankton in oil contaminated water bodies are generally greatly reduced (O'Brien and Dixon 1976). However, studies have also shown that algae can grow and survive in the presence of toxic compounds as a result of physiological adaption supported by modification of gene expression (Fogg, 2001; Lopez-Rodas et al., 2009; Oberholster et al., 2010). When the concentration of toxic elements exceeds levels of physio-logical adaptation, survival depends almost exclusively on adaptive evolution driven by mutations that confer resistance (Belfore and Anderson 2001).

Despite the large amount of literature available on crude oil and marine phytoplankton, very little is known about freshwater wetland phytoplankton species specific changes in photosynthetic activity, primary production and biomass accumulation after exposure to different concentrations of crude oil. The recent emphasis on wetland protection has created an urgent need to evaluate anthropogenic impacts on the biological integrity of wetlands, since wetland numbers are declining worldwide. Wetlands in particular act as natural biochemical 'cleansers' in the hydrological landscape and are able to break down a variety of chemical pollutants (Mitch and Gosselink, 2000).

However, data on the response of freshwater wetland biota, especially phytoplankton in relationship with crude oil pollution is sparse, given that only physico-chemical stressors are normally monitored after an oil spill.

For the purpose of this study, we conducted laboratory assays to examine the effect of different concentrations of crude oil on three, common freshwater wetland phytoplankton species that are found in the Vlakfontein depression wetland, South Africa.

The objective of this task was to establish and compare how the different species responded to various concentrations of crude oil in order to infer how wetland phytoplankton, exposed to crude oil would potentially respond, and what risk this could pose to depression wetlands if a crude oil spill occur.

The biological endpoints investigated included: chlorophyll *a* and *b* which partly functioned as surrogate for biomass, *β*-carotene to serve as stress indica-tor, gross productivity, and alkaline phosphatase activity.

MATERIALS AND METHODS

Sampling and Isolation of Phytoplankton from the Vlakfontein Depression Wetland

Phytoplankton was collected with a scoop bottle sampler (100 ml) in the littoral zone of the Vlakfontein depression wetland, Mpumalanga, South Africa (Figure 1), sedimented in a Sedgewich-Rafter counting chamber and analysed under an inverted microscope at 1200 × magnification, using the strip-count method (APHA 1992). The dominant algae were identified according to Truter (1987), Wehr and Sheath (2003), and Van Vuuren et al. (2006), and were isolated and purified.

Phytoplankton Isolation and Maintenance

Laboratory experiments were performed using four phytoplankton species of which three species belong to the green algae group chlorophyta and one to the blue green algae group cyanophyta.

The algal strain used as positive control in this study was: *Chlorella vulgaris* Beijerinck obtained from the American Tissue Culture Collection (ATCC, #30581). From the literature the latter phytoplankton species is known to increase in biomass under crude oil exposure. The three phytoplankton species collected and isolated from the Vlakfontein depression wetland were: (1) the planktonic microalgae *Chlorella protothecoides*, (2) the benthic cyano-bacteria *Chrococcus dispersus* and (3) the filamentous benthic green macro algae *Oedogonium croasdaleae*. The isolated and purified algae were excised under sterile conditions from agar plates containing different growth media (i.e., culture broth and BG-11), using a micro spade under a dissecting micro-scope. The axenic *Chlorella* spp. and *Oedogonium croasdaleae* strains were transferred to 300 ml of liquid algae culture broth medium (Sigma-Aldrich Chemie GmbH, Switzerland), while *Chrococcus dispersus* was cultured in BG-11 liquid medium (as described in ATCC: medium 616).

Figure 1. The Vlakfontein depression wetland from which freshwater algae were collected for isolation and culturing. This type of wetland is especially venerable to pollution, due to an inward draining flow pattern with no outflow.

The selected algae were cultured at 24 ± 3 °C with a light intensity of approximately 100 μmol/m^2/s, and under natural illumination (i.e., 12h:12h light:dark regime). To verify if the algae cultures were axenic, a compound microscope at 1250 x magnification was used to examine the different cultures every three days, and if the cultures were not axenic, the isolation procedure was repeated. Cultures were maintained axenic in the mid-log exponential growth phase by serial transfers of small inoculums (1 ml) to new culture medium (50 ml) every 15 days.

Crude Oil Exposures

The four phytoplankton strains were exposed to Iranian crude oil collected from the Klippoortjie underground oil bunker, in South Africa during January 2014. The treatment groups included: algae culture broth (Sigma-Aldrich, GE) supplemented with 0 mg.l^{-1}, 25 mg.l^{-1}, 250 mg.l^{-1} and 2500 mg.l^{-1} of crude oil. A further negative control containing the aforementioned crude oil concentrations without phytoplankton was also included. The experiment consisted of a 144 h static exposure, performed in triplicate using glass vessels containing 200 ml of liquid, and maintained at 24 ± 3 °C. Phytoplankton responses to crude oil was evaluated under natural illumi-nation (i.e., 12h:12h light:dark regime) and in darkness in order to calculate gross primary productivity (Adam 1990).

The "dark" exposure was performed using amber glass bottles and was conducted at the same time as the "light" exposure per phytoplankton species.

In order to adequately mix the crude oil and algae culture medium, both ingredients were sonicated (15 pulses of 2 seconds each, at 60 watts and at a frequency of 16 kHz) in 50 ml polypropylene tubes on ice using an ultrasound homogenizer (Omni-ruptor 400, Omni International Inc., US). The (sonicated) crude oil diluents were prepared separately for each individual exposure container. The *C. vulgaris*, *C. protothecoides* and *C. dispersus* experimental cultures were inoculated with 1 x 10^6 cells from mid-log exponentially growing cultures. After every 48 hours, 1 ml of algae containing culture broth was removed from each exposure vessel for chlorophyll and protein extraction purposes. The *O. croasdaleae* treatments were inoculated with 0.01 g of algae (dry mass) weighed individually.

a

b

Figure 2. Experimental setup using the dark bottle, light bottle method (Adams, 1990). Figure 1A was at the start of the experiment (time 0 h) while Figure 1B was at the end of the experiment (time 144 h).

Unlike the other species, a separate set of vessels had to be included per time point investigated (i.e., 0 h, 48 h, 96 h and 144 h), because the entire fila-mentous algae mat had to be removed for the chlorophyll and protein extrac-tion event.

Chlorophyll *a*, *b* and *β*-Carotene Quantification

Chlorophyll (Chl) *a*, *b* as well as *β*-carotene concentrations were determined for both phytoplankton and control samples as described by Dere et al. (1998) after 0, 48, 96 and 144 hours of exposure.

In short, 1 ml of the *C. vulgaris*, *C. protothecoides* and *C. dispersus* containing medium was extracted and centrifuged at 13000 x g for 2 minutes, after which the supernatant was removed. Conversely, the total *O. croasdaleae* filamentous mat was removed using forceps, reconstituted in 200 µl colori-metric assay buffer (100 mM NaCl, 100 mM Tris-Cl (pH 9.5), 50 mM $MgCl_2$ and 1 % Tween 20 (v/v)), and sonicated with an Omni-ruptor 400 homogenizer (10x 2 second pulses at 30 watt).

The filaments were subsequently centrifuged at 13000 x g for 2 minutes, after which the supernatant was removed and stored to determine alkaline phosphatase activity. The *O. croasdaleae* filaments and *C. vulgaris*, *C. proto-thecoides* and *C. dispersus* cells were reconstituted in 250 µl of methanol and incubated on a high intensity shaker for 10 min at room temperature.

Absorbance was subsequently measured at 450 nm, 655 nm and 665 nm using a spectrophotometer (Power Wave HT, BioTek, US) in order to calcula-te Chl *a*, *b* and *β*-carotene concentrations (Dere et al., 1998).

Alkaline Phosphatase (ALP) Activity Assay

After 144 h of exposure, 1 ml of the culture broth containing *C. vulgaris*, *C. protothecoides* and *C. dispersus* was extracted and centrifuged at 13000 x g for 2 minutes. The supernatant was removed and samples were stored at -80°C. The phytoplankton cells were subsequently washed with ice cold sterile-filtered PBS and centrifuged at 13000 x g for 2 minutes. Washed samples were reconstituted in 200 µl of alkaline phosphatase assay buffer, sonicated (applying 10x 2 second pulses at 30 watt) using an ultrasound homogenizer (Omni-ruptor 400, Omni International Inc., US), and centrifuged at 13 000 x g, 4 °C for 2 min.

The supernatants were used to determine the enzyme activity of *C. vulgaris*, *C. protothecoides* and *C. dispersus*, whereas the supernatants collected from the sonicated *O. croasdaleae* (see previous section on Chloro-phyll *a*, *b* and *β*-carotene quantification) was used as source for enzyme activity in the latter species.

Protein concentrations were measured per sample using a nano-drop (Thermo Scientific, US) in order to normalize ALP activity. Alkaline phospha-tase activity was determined using an ALP colorimetric assay kit according to the manufacturer's instructions (Biovision, US). The method involves the incubation of 80 µl of algae homogenate (containing 10 µg of protein per assay well or less, if 80 µl did not contain the amount) with 80 mM Tris-HCl buffer (pH 10.4), 50 µl of *p*-nitrophenyl phosphate (*p*NPP, 5 mM) at room temperature for 60 min. The reaction was terminated by adding 20 µl of stop solution (2 mM NaOH) after which absorbance was measured at 405 nm. Alkaline phosphatase activity was subsequently quantified in $\mu mol.ml^{-1}.min^{-1}$ *ρ*NPP for each sample relative to a *ρ*NPP standard curve.

Primary Gross Productivity

Gross primary productivity for each selected concentration of crude oil as well as the controls was determined by making use of the light and dark bottle method as described by Adam (1990), with slight modification. Dissolved oxygen concentrations were measured at the start and end of algae exposures using a DO700 multi meter (Extech, US). These measurements were used to calculate the gross productivity, based on a modification of the Winkler-azide test and expressed according to the following equations:

$$\text{Respiration} = \frac{[\text{initial oxygen}\left(\frac{\text{mg}}{\text{L}}\right) - \text{dark bottle}\left(\frac{\text{mg}}{\text{L}}\right)]}{\text{Time in hours}} \quad (1)$$

$$\text{Net productivity} = \frac{[\text{Light bottle}\left(\frac{\text{mg}}{\text{L}}\right) - \text{initial oxygen}\left(\frac{\text{mg}}{\text{L}}\right)]}{\text{Time in hours}} \quad (2)$$

Gross productivity = Net productivity – Respiration equation (2) - equation (1)

$$= \frac{(\text{light bottle} - \text{inital oxygen}) - (\text{initial oxygen} - \text{dark bottle})]}{\text{Time in hours}}$$

$$= \frac{[\text{light bottle} - \text{dark bottle}]}{\text{hours}} \quad (3)$$

Water Chemistry

The pH and electrical conductivity of each exposure vessel was recorded at 48 hour intervals throughout the 144 h exposure period.

Statistical Analysis

Normality and homogeneity of variance of data was assessed using the Shapiro-Wilk's W test, normal probability plots of residuals and Levene's test respectively. Boxcox or rank transformations were applied to non-parametric data in order to meet ANOVA assumptions. Temporal trends in pigment concentrations were evaluated using generalized linear mixed model repeated measures analysis of variance in combination with the Fisher's LSD *post hoc* test (Variance Estimation and Precision [VEPAC] module, Statistica 12, Statsoft, US). Interspecific variation of pigment concentrations in response to crude oil were evaluated using factorial ANOVA (Statistica 12, Statsoft, US). Principal component analyses were performed using XLSTAT (Addinsoft, US). Probability values (P) of less than 0.05 were accepted as significant.

RESULTS

Chlorophyll *a*, *b* and *β*-Carotene Concentrations

Chlorophyll *a*, *b* concentrations varied significantly as a function of both time and crude oil treatment in *C. vulgaris, C. protothecoides* and *C. dispersus* across the 144 h exposure period (Figure 3, Figure 4, Table 1). The temporal variation of Chl *a* and Chl *b* was most pronounced in *C. protothecoides* (Figure 3). In particular, the control and three crude oil treatments had significantly higher concentrations of both pigments after 144 h relative to the 0 h, 48 h and 96 h time points (Figure 3, $P < 0.05$, Fisher's LSD *post hoc*), accept for the 2500 mg.l^{-1} treatment in which chl *a* concentrations did not vary significantly between the 96 h and 144 h time points, albeit a step-wise temporal increase (Figure 3).

Similar to Chl *a* and Chl *b*¸ *β*-carotene concentrations also varied signifi-cantly over time in *C. vulgaris, C. protothecoides* and *C. dispersus*, whereas crude oil dose dependent variation in these pigments were only significant in *C. vulgaris* and *C. dispersus* (Table 1). The selected crude oil concentrations and time interaction showed a significant variance in the *β*-carotene concentrations *of C. protothecoides* (Table 1), suggesting a time-dependent variability in response to crude oil treatment as is evident in Figure 3.

The temporal trend of *β*-carotene levels corresponded to that of Chl *a* and Chl *b*, being most pronounced in *C. protothecoides* (Figure 3), and *β*-carotene was significantly higher after 144 h relative to the other time points investigated (Figure 3, $P < 0.05$, Fisher's LSD *post hoc*). In *C. protothecoides* (Figure 3), the 25 mg.l^{-1} crude oil treatment furthermore resulted in the most pronounced increases in Chl *a*, *b* and *β*-carotene concentrations, and all three compounds were significantly higher than the control treatment after 144 h (Figure 3, $P < 0.05$, Fisher's LSD *post hoc*). In contrast to *C. vulgarus* and *C. protothecoides*, no temporal increase in Chl *a*, *b* or *β*-carotene concentrations were observed in the *C. dispersus* control group (Figure 4). However, Chl *a*, *b* and *β*-carotene concentrations were significantly higher in both the 250 mg.l^{-1} and 2500 mg.l^{-1} crude oil treatments relative to the control after 144 h of exposure (Figure 4, $P < 0.05$, Fisher's LSD *post hoc*). The Chl *a* concentrations in the *O. croasdaleae* control treatment increased significantly after 144 h (Figure 4, $P < 0.05$, Fisher's LSD *post hoc*). Although a similar temporal trend was evident for Chl *b* and *β*-carotene concentrations, these differences were not significant (Figure 4, $P > 0.05$, Fisher's LSD *post hoc*). It was evident from the data that the selected crude oil treatment was a significant source of variance in Chl *a*, Chl *b* and *β*-carotene concentrations in the algae *O. croasdaleae* (Table 1), in comparison to other phytoplankton species investigated in this study. In the case of *O. croasdaleae* all three pigments decreased over time in response to crude oil exposure (Figure 3 and 4). In particular, Chl *a* decreased significantly from 0 h to 144 h in both the 25 mg.l^{-1} and 2500 mg.l^{-1} treatments (Figure 3, $P < 0.05$, Fisher's LSD *post hoc*). A similar decrease associated with *O. croasdaleae* exposure to 25 mg.l^{-1} and 2500 mg.l^{-1} was evident in Chl *b* and *β*-carotene concentrations (Figure 4), yet these differences were only significant in the case of *β*-carotene ($P < 0.05$, Fisher's LSD *post hoc*).

Moreover, Chl *a*, *b* and *β*-carotene concentrations were significantly lower in response to 25 mg.l^{-1} and 2500 mg. l^{-1} crude oil than in the control treatment after 144 h of exposure (Figure 4, $P < 0.05$, Fisher's LSD *post hoc*).

Chlorophyll *a*:*b* Ratios

Chlorophyll *a*:*b* ratios were recorded after 144 h of exposure for all four phytoplankton species (Figure 5), and varied significantly among crude oil treatments in both *C. protothecoides* ($F_{3,8} = 8.21$, $P = 0.008$) and *O. croasdaleae* ($F_{3,8} = 4.42$, $P = 0.041$). The only significant pairwise differences in Chl *a*:*b* ratios were however in *C. protothecoides*, where the ratio of the 2500 $mg.l^{-1}$ treatment was significantly lower than both the 0 $mg.l^{-1}$ and 25 $mg.l^{-1}$ treatments ($P < 0.05$, Unequal *N* Tukey Spjotvoll/Stoline HSD *post hoc*, Figure 5).

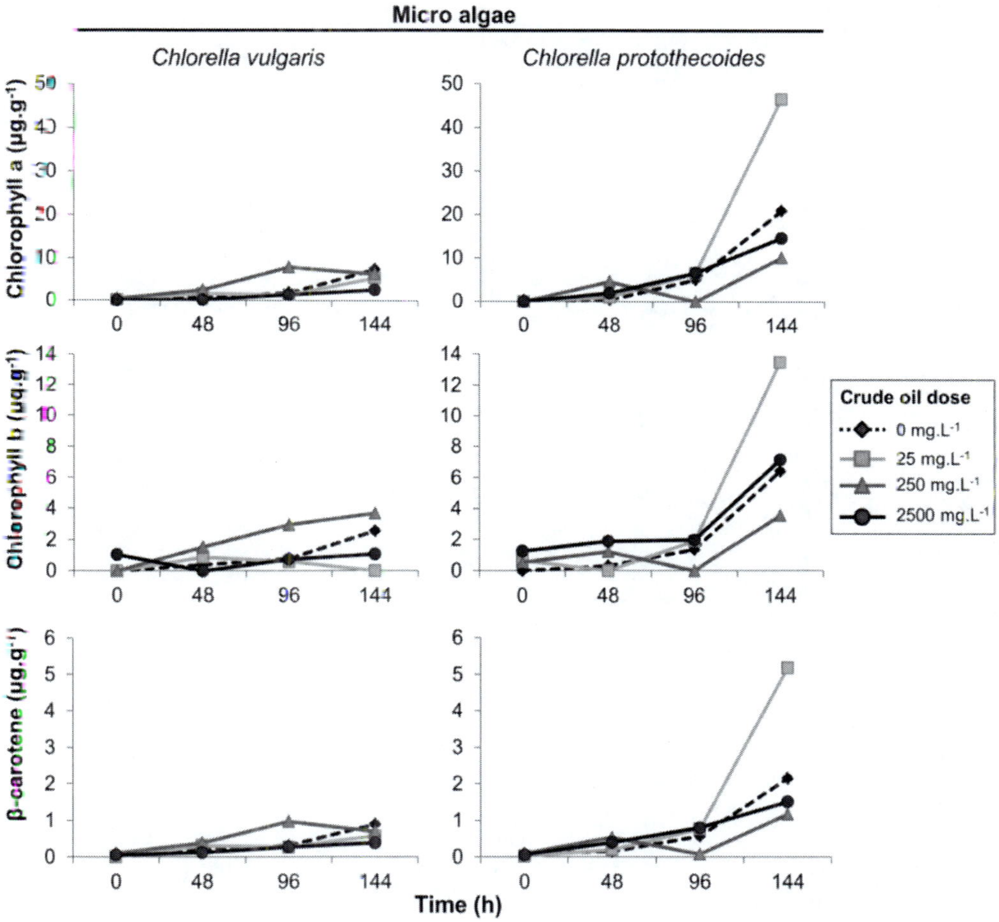

Figure 3. Temporal changes in chlorophyll *a*, chlorophyll *b* and *β*-carotene concentrations in response to different concentrations of crude oil. Each point represents the mean of three replicates per crude oil dose, per algae species.

Gross Productivity

Gross productivity did not vary significantly among the different crude oil concentrations in any of the phytoplankton species investigated (Figure 6). There was however a trend of increased productivity in response to crude oil exposure in *C. vulgaris*, *C. protothecoides* and *O. croasdaleae* (Figure 6).

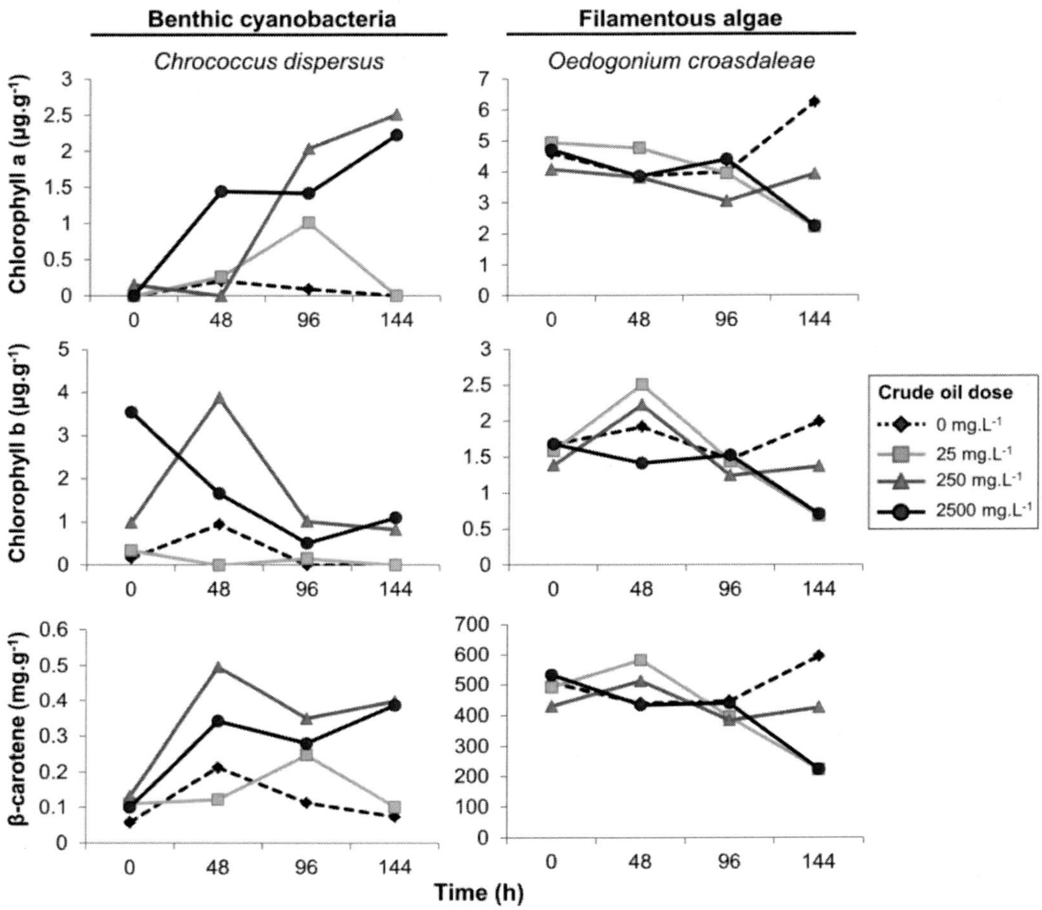

Figure 4. Temporal changes in chlorophyll *a*, chlorophyll *b* and *β*-carotene concentra-tions in response to different concentrations of crude oil. Each point represents the mean of three replicates per crude oil dose, per phytoplankton species.

Alkaline Phosphatase

Alkaline phosphatase activity (APA) could not be detected for *C. vulgaris* or *C. protothecoides* in any of the crude oil treatments after 144 h of exposure, whereas APA for *C. dispersus* could only be detected in the 25 mg.l^{-1} crude oil treatment (Table 5).

Table 1. Variation in chlorophyll *a*, chlorophyll *b* and *β*-carotene concentrations among algae species as a function of time and crude oil treatment. Variance was evaluated using generalized linear mixed model repeated measures analysis of variance for *C. vulgaris*, *C. protothecoides* and *C. dispersus*, whereas generalized linear model analysis of variance was applied for *O. croasdaleae* due to the alternative experimental design for this species. Numerator and denominator degrees of freedom (*df*) are indicated

Compound	Source of variation	*C. vulgaris*			*C. protothecoides*			*C. dispersus*			*O. croasdaleae*		
		df	*F*	*P*	*df*	*F*	*P*	*df*	*F*	*P*	*df*	*F*	*P*
	Time	3,24	57.35	< 0.001***	3,24	169.52	< 0.001***	3,24	28.51	< 0.001***	3,32	1.740	0.179
Chlorophyll *a*	Dose	3,8	17.68	< 0.001***	3,8	8.43	0.007**	3,8	18.23	< 0.001***	3,32	3.431	0.029*
	Time*Dose	9,24	1.65	0.158	9,24	21.42	< 0.001***	9,24	7.14	< 0.001***	9,32	2.524	0.026*
	Time	3,24	8.12	< 0.001***	3,24	22.10	< 0.001***	3,24	5.44	0.005**	3,32	3.912	0.017*
Chlorophyll *b*	Dose	3,8	7.89	0.009**	3,8	3.10	0.089	3,8	34.34	< 0.001***	3,32	4.975	0.006**
	Time*Dose	9,24	3.34	0.009**	9,24	4.87	0.001**	9,24	4.52	< 0.001***	9,32	1.540	0.176
	Time	3,24	62.24	< 0.001***	3,24	257.91	< 0.001***	3,24	32.55	< 0.001***	3,32	2.394	0.087
β-carotene	Dose	3,8	19.70	< 0.001***	3,8	2.28	0.156	3,8	16.55	< 0.001***	3,32	3.144	0.039
	Time*Dose	9,24	8.30	< 0.001***	9,24	21.30	< 0.001***	9,24	6.73	< 0.001***	9,32	1.840	0.099
	Time	3,14	8.58	0.002**	3,16	13.96	< 0.001***	3,13	9.52	0.001**	3,32	12.49	< 0.001***
Chl *a*:*b* ratio	Dose	3,8	0.24	0.863	3,8	2.17	0.170	3,8	0.40	0.755	3,32	2.31	0.095
	Time*Dose	6,14	0.47	0.816	8,16	2.17	0.090	5,13	0.81	0.562	9,32	0.92	0.519

Significant differences are indicated with asterisks.

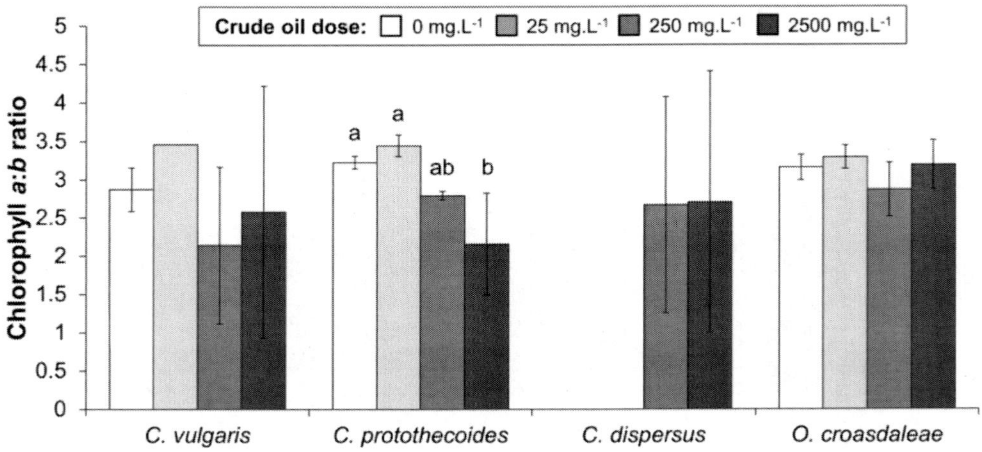

Figure 5. Chlorophyll *a:b* ratio (Mean ± SD) in response to crude oil dose after 144 h of exposure. Dissimilar characters indicate significant differences ($P < 0.05$, Unequal *N* HSD *post hoc*).

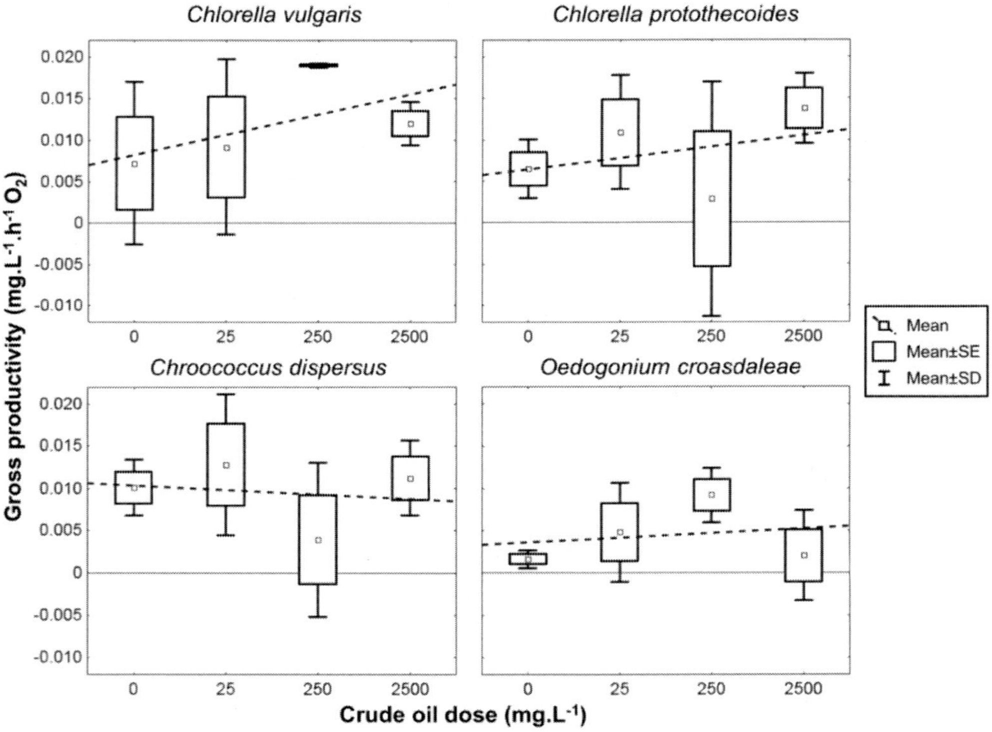

Figure 6. Gross productivity ($mg.l^{-1}.h^{-1}$) of *C. vulgaris*, *C. protothecoides*, *C. dispersus* and *O. croasdaleae* in response to a series of crude oil dilutions.

Conversely, *O. croasdaleae* had relatively consistent APA activity among the different crude oil treatment concentrations, and these activities did not vary significantly (Table 5).

Table 5. Alkaline phosphatase activity (µmol.ml^{-1}.min^{-1} *ρ*NPP) of *Chlorella vulgaris*, *Chlorella protothecoides*, *Chrococcus disperses* and *Oedogonium croasdaleae* in response to a series of crude oil dilutions after 144 hours of exposure. Each value represents the mean ± standard deviation of three replicate treatments

Crude oil dose	Alkaline phosphatase activity (µmol.ml^{-1}.min^{-1} *ρ*NPP)			
	C. vulgaris	*C. protothecoides*	*C. dispersus*	*O. croasdaleae*
0 mg.L^{-1}	ND	ND	ND	0.66 ± 0.29
25 mg.L^{-1}	ND	ND	0.03 ± 0.05	0.69 ± 0.22
250 mg.L^{-1}	ND	ND	ND	0.46 ± 0.41
2500 mg.L^{-1}	ND	ND	ND	0.63 ± 0.06

ND: Not detectable.

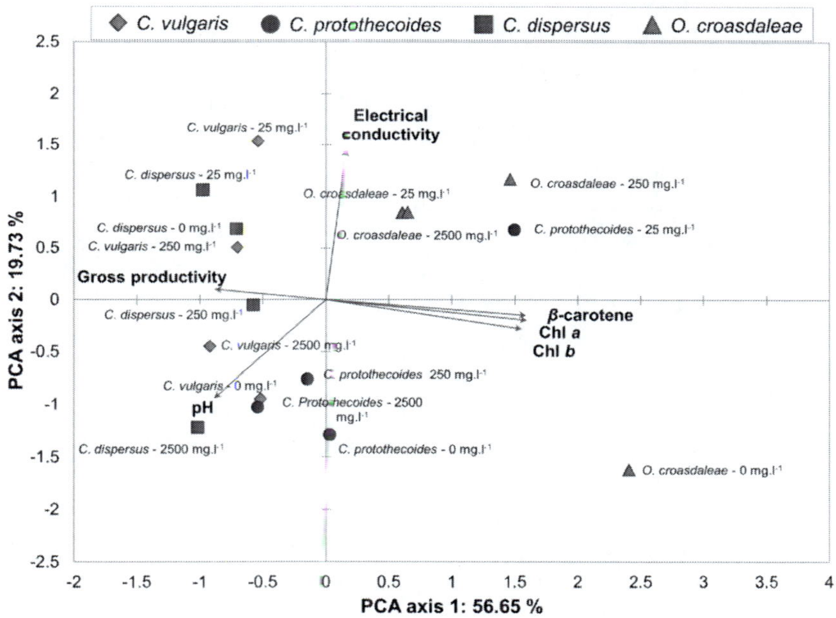

Figure 7. A principal component analysis biplot indicating the associations between crude oil dose, chlorophyll *a* (Chl *a*), Chl *b*, *β*-carotene, gross productivity, electrical conductivity (EC) and pH among the four phytoplankton species, *Chlorella vulgaris*, *Chlorella protothecoides*, *Chrococcus disperses* and *Oedogonium croasdaleae*.

Chlorophyll *a*, *b*, *β*-Carotene, Gross Productivity, pH and Electrical Conductivity

From the principal component analysis it was evident that no distinct crude oil treatment or phytoplankton species dependent pattern was distin-guishable in light of Chl *a*, Chl *b*, β-carotene, gross productivity or the pH and electrical conductivity.

Discussion

Studies by Ansari, (1997) and El-Sheekh et al. (2000) indicated that crude oil is less toxic to phytoplankton than refined oils, since refined oils are en-riched with different aromatic compounds e.g., benzene and naphthalene.

According to Huang et al. (2011) high concentrations of the water soluble fraction of oil inhibit phytoplankton growth and decreases cell density. However, it is known that certain species of phytoplankton may use oil hydrocarbons as an additional carbon source (Tkalich et al., 1999). A study by O'Brien and Dixon (1976) showed that elevated growth levels of *Chlorella vulgaris* under heavy oil pollution was due to this species' ability to make use of certain oil hydrocarbons. Furthermore, Huang et al. (2011) reported that at low concentrations of crude oil contamination, oxygen free radicals present in the weathering and photo-activated oil may promote the production of super-oxide dismutase (SOD), which may be one of the causes for higher resistance to crude oil contamination in the more tolerant phytoplankton species. On the contrary, other authors have found that at very low crude oil concentrations the growth of the exposed phytoplankton was severely inhibited by petroleum hydrocarbons while sub-lethal levels of oil can induce stress in estuarine phytoplankton communities (Ansari 1997; Gilde and Pinckney, 2012; Özhan et al., 2014). The responses to crude oil are typically species dependent (Gilde and Pinckney, 2012; El Seekh et al., 2013; Özhan et al., 2014), as has been observed in the current study. In particular, the varied response as a function of phytoplankton species and crude oil treatment can be seen on Figure 7, where no clear crude oil dose or species dependent pattern is distinguishable in light of Chl *a*, Chl *b*, *β*-carotene, gross productivity or the pH and electrical conductivity of the media.

Individual phytoplankton cells were the most sensitive to oil contamina-tion when steady state growth was achieved (Huang et al., 2011).

Nevertheless, if the concentration of oil exceeds the cell threshold, it may prevent the cell from growing. The reduction of cell growth may also be contributed to the decrease in cellular ATP content due to the impairment of the photosynthetic apparatus (Oberholster et al., 2010; Perez et al., 2010). O'Brien and Dixon, (1976) showed that crude oil acts as a physical barrier for the diffusion of CO_2 across phytoplankton cell walls and the absorption of nutrients, resulting in the decrease of chlorophyll *a* and reduction of primary productivity. Observations by Jiang et al. (2010) and Pelletier et al. (2006) indicate that lipophilic hydrocarbons of the water soluble fraction bioaccumu-late in marine phytoplankton and can have a direct toxic effect or suppress their photosynthesis by reducing the primary photochemical yield. A study by Singh and Gaur (1990) showed that crude oil acts primarily on PSII, probably through direct interaction with the electron transport system. It is a known fact that chlorophyll *a* concentrations are often used as an indirect measure of phytoplankton productivity and therefore a reduction in chlorophyll *a* is also a decrease in productivity (Echeveste et al., 2010). Nonetheless, the loss of photosynthetic pigments, are evidently a reversible process in sublethal envi-ronments (O'Brien and Dixon, 1976). According to Romero-Lopez et al. (2012), phytoplankton cell exposure to PAHs in particular, alters and reduces expression of genes in charge of photosynthetic pigment production which include chlorophylls and carotenoids.

From the study by Romero-Lopez et al. (2012) it was evident that some cyanobacteria and microalgae can adapt to crude oil pollution rather quickly by single gene mutation and are also able to proliferate under low oil concentrations as a result of physiological acclimatization. In addition to the reduction of the electron transport capacity, the water soluble fraction also decreases the activity of the release of oxygen centers (Jiang et al. 2010).

Under these conditions, where the production of organic molecules was inhibited, respiration enhances to satisfy the energy demand of the marine phytoplankton cells. Furthermore, petroleum also penetrates phytoplankton cell mitochondria and damages their membranes, resulting in the loss of the ability to respirate aerobically (O'Brien and Dixon, 1976).

The Effects of Crude Oil on the Phytoplankton Assemblage of a Freshwater Wetland

The results from the current study support the view that the variability in the resistance to oil pollution of distinct phytoplankton species may lead to a non-normal phytoplankton assemblage, with the more tolerant species becoming the dominant group (Oberholster et al., 2010).

Furthermore, the aquatic wetland system may be affected when zooplank-ton communities are altered due to selected feeding on phytoplankton causing a bottom up effect (Perez et al. 2010). The shift in phytoplankton communities can alter the structure, stability and function of a wetland aquatic system.

Additionally, the die off of crude oil intolerant species, may release important carbon and nutrient sources for bacteria, which can cause a transi-tion from an phytoplankton based foodweb to a microbial dominated foodweb.

Conclusion

In this study the impact of increasing crude oil concentrations on the proliferation and productivity of three algal species were investigated. The results indicate that there was variability in the susceptibility of distinct phytoplankton species to the selected crude oil concentrations. In general, exposure to crude oil inhibits phytoplankton photosynthetic efficiency, but from the study it was evident that some species can tolerate crude oil or even be stimulated to grow. The benthic filamentous alga *O. croasdaleae* was however unable to adapt to the presence of crude oil, suggesting that an oil spill may pose a potential risk to the benthic aquatic food chain of depression wetlands.

References

Adam, V. D., 1990. *Water and wastewater examination manual*, Lewis Publishers, Michigan.

Ansari, Z. A., Sadanha, M. C., Rajkumar, R., 1997. Effects of petroleum hydrocarbons on the growth of a microalga, *Isochrysis* sp. (Crysophyta) *Indian Journal of marine science* 26: 372-376.

APHA. 1992. *Standard Methods for the Examination of Water and Waste-water,* 18th edition. American Public Health Association, Washington, D.C.

Belfore, N. M., Anderson, S. I., 2001. Effects of contaminants on genetic patterns in aquatic organisms: a review. *Mutation research* 489: 97-122.

Dawes, C. J., 1998. *Marine Botany,* second ed. J. Wiley and Sons. Inc. New York, US.

Dere, S., Gunes, T., Sivaci, R., 1998. Spectrophotometric determination of dhlorophyll - a, b and total carotenoid contents of some algae species using different solvents. *Turkish journal of botany*. 22: 13-17.

Echeveste, P., Dachs, J., Berrojalbiz, N., Agusti, S., Dachs, J., 2010. Cell size dependent toxicity thresholds of polycyclic aromatic hydrocarbons to natural and cultured phytoplankton populations. *Environmental pollution* 158: 299-307.

El-Sheekh, M., El-Naggar, A. H., Osman, M. E. H., Haieder, A., 2000. Com-parative studies on the green algae *Chlorella homosphaera* and *Chlorella vulgaris* with respect to oil pollution in the river Nile. *Water, Air and Soil Pollution* 124: 187-204.

El-Sheekh, M., Hamouda, R. A., Nizam, A. A., 2013. Biodegredation of crude oil by *Scenedemiodus obliquus* and *Chlorella vulgaris* under heterotrophic conditions. *International journal of biodeterioration and biodegradation* 82: 67-72.

Fogg, G. E., 2001. Algal adaptation to stress, some general remarks. In: Rai, L. C. and Gauf, J. P. (Eds.). *Algal adaption to environmental stresses, physiological, biochemical and molecular mechanisms*. Springer, Berlin, Germany, pp. 1-20.

Gilde, K., Pinckney, J. L. 2012. Sublethal effects of crude oil on the commu-nity structure of estuarine phytoplankton. *Estuaries and coasts* 35: 853-861.

Huang, Y. J., Jiang, Z. B., Zeng, J. N., Che, Q. Z., Zhao, Y. Q., Liao, Y. B., Shou, L., Xu, X. Q. 2011. The chronic effects of oil pollution on marine phytoplankton in a subtropical bay, China. *Environmental monitoring and assessment* 176: 517-530.

Jiang, Z., Y. Huang, X. Xu, Y. Liao, L. Shou., J. Liu, Q. Chen, J. Zeng, 2010. Advance in the toxic effects of petroleum water accommodated fraction on marine plankton. *Acta Ecologica Sinica* 30: 8-15.

Lopez-Rodas, V., Carrera-Martinez, D., Salgado, E., Mateos-Sanz, A., Baez, J. C., Costas, E., 2009. A facinatingexample of microalgal adaption to extreme crude oil contamination in a natural spill in Arroyo Minero, Rio Negro, Argentina. *Anales De La Real Academia De Farmacia*. 75: 883-899.

Mitsch, W. J., Gosselink, J. G., 2000. *Wetlands*. Van Nostrand Reinhold, New York.

O'Brien, J. P., Dixon, P. S., 1976. The effects of oil and oil components on algae: A review. *British phycological journal* 11: 115-142.

Oberholster, P. J., Blaise, C. B. J., Botha, A.-M., 2010. Phytobenthos and phytoplankton community changes upon exposure to a sunflower oil spill in a South African protected freshwater wetland. *Ecotoxicology* 19: 1426-1439.

Özhan, K., Miles, S., Gao, H., and Bargu, S., 2014. Relative Phytoplankton growth responses to physically and chemically dispersed South Louisiana sweet crude oil. *Environmental monitoring and assessment.* In press.

Pelletier, E., Sargian, P., Sargian, Payet, J., Demers, S., 2006. Ecotoxicological effects of combined UVB and Organic contaminants in coastal waters: Areview. *Photochemistry and Photobiology* 82: 981-987.

Perez, P. Fernandez, E., Beira, R. 2010. Fuel toxicity on *Isochrysis galbana* and a coastal phytoplankton assemblage: growth rate vs. variable fluore-scence. *Ecotoxicology and environmental* safety 73: 254-261.

Romero-Lopez, J. V., Lopez-Rodas, V., Costas, E., 2012. Estimating the capability of microalgae to physiological acclimatization and genetic adaptation to petroleum and diesel oil contamination. *Aquatic toxicology* 124: 227-237.

Singh, A. K., Gaur, J. P., 1990. Effects of petroleum oils and their paraffinic, asphaltic, and aromatic fractions on photosynthesis and respiration of microalgae. *Ecotoxicology and environmental safety* 19: 8-16.

Truter, E., 1987: An aid to the identification of the dominant and commonly occuring genera of algae observed in some South African impoundments. *Technical Report TR 135*. Department of Water Affairs. – Hydrological Research Institute. Pretoria. 101 pp.

Tkalich, P., Huda, K., Gin, K. 1999. A Multiphase model of oil spill dynamics. In: *Proceedings of the IAHR Congress*, US, pp. 22-27.

Van Vuuren, S., Taylor, J. C., Gerber, A., Van Ginkel, C., *2006: Easy identifi-cation of the most common freshwater algae.* – North-West University and Department of Water Affairs and Forestry, Pretoria, South Africa, pp. 1-200.

Wehr, J. D., Sheath, R. G., *2001: Freshwater Algae of North America, Ecolo-gy and Classification.* – Academic Press, San Diego, US. pp. 775 – 804.

INDEX

A

B

D

E

F

G

H

M

N

O

P

Q

R

S

T

U

V

W

Y

Z